D1741987

1 MONTH OF
FREE
READING

at
www.ForgottenBooks.com

By purchasing this book you are eligible for one month membership to ForgottenBooks.com, giving you unlimited access to our entire collection of over 1,000,000 titles via our web site and mobile apps.

To claim your free month visit:
www.forgottenbooks.com/free1112279

ISBN 978-0-331-36506-1
PIBN 11112279

Standard Reference Materials:

Critical Current Measurements on an NbTi Superconducting Wire Standard Reference Material

L. F. Goodrich[1]
D. F. Vecchia[2]
E. S. Pittman[1]
J. W. Ekin[1]
A. F. Clark[1]

[1] Center for Electronics and Electrical Engineering
National Engineering Laboratory
National Bureau of Standards
Boulder, CO 80303

[2] Center for Applied Mathematics
National Engineering Laboratory
National Bureau of Standards
Boulder, CO 80303

Sponsored by:
Office of Standard Reference Materials
National Measurement Laboratory
National Bureau of Standards
Gaithersburg, MD 20899
and
Office of Fusion Energy
and
Office of High Energy Physics
Department of Energy
Washington, DC 20545

NBS *Special Publication*

U.S. DEPARTMENT OF COMMERCE, Malcolm Baldrige, Secretary

NATIONAL BUREAU OF STANDARDS, Ernest Ambler, Director

Issued September 1984

Library of Congress Catalog Card Number: 84-601108

National Bureau of Standards Special Publication 260-91
Natl. Bur. Stand. (U.S.), Spec. Publ. 260-91, 67 pages (Sept. 1984)
CODEN: XNBSAV

U.S. GOVERNMENT PRINTING OFFICE
WASHINGTON: 1984

Standard Reference Materials (SRM's) as defined by the National Bureau of Standards are "well-characterized materials, produced in quantity, that calibrate a measurement system to assure compatibility of measurement in the Nation." SRM's are widely used as primary standards in many diverse fields of science, industry, and technology, both within the United States and throughout the world. For many of the Nation's scientists and technologists it is of more than passing interest to know the measurements obtained and methods used by the analytical community when analyzing SRM's. An NBS series of papers, of which this publication is a member, called the NBS Special Publication - 260 Series is reserved for this purpose.

This 260 Series is dedicated to the dissemination of elemental concentration data for NBS biological, geological, and environmental SRM's. More information will be found in this 260 than is generally found in NBS Certificate of Analysis. This 260 enables the user of these SRM's to assess the validity of data not available in the Certificate of Analysis. We hope that this 260 will provide sufficient additional information so that new applications of these SRM's may be sought and found.

Inquiries concerning the technical content of this compilation should be directed to the authors. Other questions concerned with the availability, delivery, price of specific SRM's should be addressed to:

> Office of Standard Reference Materials
> National Bureau of Standards
> Gaithersburg, MD 20899

> > Stanley D. Rasberry, Chief
> > Office of Standard Reference Materials

Catalog of NBS Standard Reference Materials (1981-83 edition), R. W. Seward, ed., NBS Spec. Publ. 260 (November 1981).

Michaelis, R. E., and Wyman, L. L. Standard Reference Materials: Preparation of White Cast Iron Spectrochemical Standards. NBS Misc. Publ. 260-1 (June 1964). COM74-11061**

Michaelis, R. E., Wyman, L. L., and Flitsch, R., Standard Reference Materials: Preparation of NBS Copper-Base Spectrochemical Standards. NBS Misc. Publ. 260-2 (October 1964). COM74-11063**

Michaelis, R. E., Yakowitz, H., and Moore, G. A., Standard Reference Materials: Metallographic. Characterization of an NBS Spectrometric Low-Alloy Steel Standard. NBS Misc. Publ. 260-3 (October 1964). COM74-11060**

Hague, J. L. Mears, T. W., and Michaelis, R. E., Standard Reference Materials: Sources of Information, NBS Misc. Publ. 260-4 (February 1965). COM74-11059

Alvarez, R., and Flitsch R., Standard Reference Materials: Accuracy of Solution X-Ray Spectrometric Analysis of Copper-Base Alloys. NBS Misc. Publ. 260-5 (March 1965). PB168068**

Shultz, J. I,. Standard Reference Materials: Methods for the Chemical Analysis of White Cast Iron Standards, NBS Misc. Publ. 260-6 (July 1975). COM74-11068**

Bell, R. K., Standard Reference Materials: Methods for the Chemical Analysis of NBS Copper-Base Spectrochemical Standards. NBS Misc. Publ. 260-7 (October 1965). COM74-11067**

Richmond, M.S, Standard Reference Materials: Analysis of Uranium Concentrates at the National Bureau of Standards. NBS Misc. Publ. 260-8 (December 1965). COM74-11066**

Anspach, S. C., Cavallo, L. M. Garfinkel, S. B. Hutchinson, J. M. R., and Smith, C. N., Standard Reference Materials: Half Lives of Materials Used in the Preparation of Standard Reference Materials of Nineteen Radioactive Nuclides Issued by the National Bureau of Standards NBS Misc. Publ. 260-9 (November 1965). COM74-11065**

Yakowitz, H., Vieth, D. L., Heinrich, K. F. J., and Michaelis, R. E., Standard Reference Materials: Homogeneity Characterization on NBS Spectrometric Standards II: Cartridge Brass and Low-Alloy Steel, NBS Misc. Publ. 260-10 (December 1965). COM74-11064**

Napolitano, A., and Hawkins, E. G., Standard Reference Materials: Viscosity of Standard Lead-Silica Glass, NBS Misc. Publ. 260-11 (November 1966). NBS Misc. Publ. 260-11**

Yakowitz, H., vieth, D. L., and Michaelis, R. E., Standard Reference Materials: Homogeneity Characterization of NBS Spectrometric Standards III: White Cast Iron and Stainless Steel Powder Compact, NBS Misc. Publ. 260-12 (September 1966). NBS Misc. Publ. 260-12**

Spijkerman, J. L., Snediker, D. K., Ruegg, F. C., and DeVoe, J. R., Standard Reference Materials: Mossbauer Spectroscopy Standard for the Chemical Shift of Iron Compounds, NBS Misc. Publ. 260-13 (July 1967). NBS Misc. Publ. 260-13**

Menis, O., and Sterling, J. T., Standard Reference Materials: Determination of Oxygen in Ferrous Materials - SRM 1090, 1091, and 1092, NBS Misc. Publ. 260-14 (September 1966). NBS Misc. Publ. 260-14**

Passaglia, E., and Shouse, P. J. Standard Reference Materials: Recommended Method of Use of Standard Light-Sensitive Paper for Calibrating Carbon Arcs Used in Testing Textiles for Colorfastness to Light, NBS Misc. Publ. 260-15 (June 1967). (Replaced by NBS Spec. Publ. 260-41.)

Yakowitz, H., Michaelis, R. E., and Vieth, D. L., Standard Reference Materials: Homogeneity Characterization of NBS Spectrometric Standards IV: Preparation and Microprobe Characterization of W-20% MO Alloy Fabricated by Powder Metallurgical Methods, NBS Spec. Publ. 260-16 (January 1969). COM74-11062**

Catanzaro, E. J., Champion, C. E., Garner, E. L., Marinenko, G., Sappenfield, K. M., and Shields, W. R. Standard Reference Materials: Boric Acid; Isotopic and Assay Standard Reference Materials, NBS Spec. Publ. 260-17 (February 1970). Out of Print

Geller, S. B., Mantek, P.A., and Cleveland, N. G., Standard Reference Materials: Calibration of NBS Secondary Standard Magnetic Tape (Computer Amplitude Reference) Using the Reference Tape Amplitude Measurement "Process A, "NBS Spec. Publ. 260-18 (November 1969). (See NBS Spec. Publ. 260-29.)

Paule, R. C., and Mandel, J., Standard Reference Materials: Analysis of Interlaboratory Measurements on the Vapor Pressure of Gold (Certification of Standard Reference Material 745). NBS Spec. Publ. 260-19 (January 1970). PB190071**

Paule, R. C., and Mandel, J., Standard Reference Materials: Analysis of Interlaboratory Measurements on the Vapor Pressures of Cadmium and Silver, NBS Spec. Publ. 260-21 (January 1971). COM74-11359**

Yakowitz, H., Fiori, C. E., and Michaelis, R. E., Standard Reference Materials: Homogeneity Characterization of Fe-3 Si Alloy, NBS Spec. Publ. 260-22 (February 1971). COM74-11357**

Napolitano, A., and Hawkins, E. G., Standard Reference Materials: Viscosity of a Standard Borosilicate Glass, NBS Spec. Publ. 260-23 (December 1970). COM71-00157**

Sappenfield, K. M., Marineko, G., and Hague, J. L., Standard Reference Materials: Comparison of Redox Standards, NBS Spec. Publ. 260-24 (January 1972). COM72-50058**

Hicho, G. E., Yakowitz, H., Rasberry, S. D., and Michaelis, R. E., Standard Reference Materials: A Standard Reference Material Containing Nominally Four Percent Austenite, NBS Spec. Publ. 260-25 (February 1971). COM74-11356**

Martin, J. F., Standard Reference Materials: National Bureau of Standards-US Steel Corportion Joint Program for Determining Oxygen and Nitrogen in Steel, NBS Spec. Publ. 260-26 (February 1971). 85 cents* PB 81176620

Garner, E. L., Machlan, L. A., and Shields, W. R., Standard Reference Materials: Uranium Isotopic Standard Reference Materials, NBS Spec. Publ. 260-27 (April 1971). COM74-11358**

Heinrich, K. F. J., Myklebust, R. L., Rasberry, S. D., and Michaelis, R. E., Standard Reference Materials: Preparation and Evaluation of SRM's 481 and 482 Gold-Silver and Gold-Copper Alloys for Microanalysis, NBS Spec. Publ. 260-28 (August 1971). COM71-50365**

Geller, S. B., Standard Reference Materials: Calibration of NBS Secondary Standard Magnetic Tape (Computer Amplitude Reference) Using the Reference Tape Amplitude Measurement "Process A-Model 2," NBS Spec. Publ. 260-29 (June 1971). COM71-50282

Gorozhanina, R. S., Freedman, A. Y., and Shaievitch, A. B. (translated by M. C. Selby), Standard Reference Materials: Standard Samples Issued in the USSR (A Translation from the Russian). NBS Spec. Publ. 260-30 (June 1971). COM71-50283**

Hust, J. G., and Sparks, L. L., Standard Reference Materials: Thermal Conductivity of Electrolytic Iron SRM 734 from 4 to 300 K, NBS Spec. Publ. 260-31 (November 1971). COM71-50563**

Mavrodineanu, R., and Lazar, J. W., Standard Reference Materials: Standard Quartz Cuvettes, for High Accuracy Spectrophotometry, NBS Spec. Publ. 260-32 (December 1973). 55 cents* SN003-003-01213-1

Wagner, H. L., Standard Reference Materials: Comparison of Original and Supplemental SRM 705, Narrow Molecular Weight Distribution Polystyrene, NBS Spec. Publ. 260-33 (May 1972). COM72-50526**

Sparks, L. L., and Hust, J. G., Standard Reference Materials: Thermoelectric Voltage, NBS Spec. Publ. 260-34, (April 1972). COM72-50371**

Sparks, L. L., and Hust, J. G., Standard Reference Materials: Thermal Conductivity of Austenitic Stainless Steel, SRM 735 from 5 to 280 K, NBS Spec. Publ. 260-35 (April 1972.) 35 cents* COM72-50368**

Cali, J. P., Mandel, J., Moore, L. J., and Young, D. S., Standard Reference Materials: A Referee Method for the Determination of Calcium in Serum, NBS SRM 915, NBS Spec. Publ. 260-36 (May 1972). COM72-50527**

Shultz, J. I. Bell., R. K. Rains, T. C., and Menis, O., Standard Reference Materials: Methods of Analysis of NBS Clay Standards, NBS Spec. Publ. 260-37 (June 1972). COM72-50692**

Richmond, J. C., and Hsia, J. J., Standard Reference Materials: Preparation and Calibration of Standards of Spectral Specular Reflectance, NBS Spec. Publ. 260-38 (May 1972). COM72-50528**

Clark, A. F., Denson, V.A., Hust, J. G., and Powell, R. L., Standard Reference Materials The Eddy Current Decay Method for Resistivity Characterization of High-Purity Metals, NBS Spec. Publ. 260-39 (May 1972). COM72-50529**

McAdie, H. G., Garn, P.D., and Menis, O., Standard Reference Materials: Selection of Thermal Analysis Temperature Standards Through a Cooperative Study (SRM 758, 759, 760), NBS Spec. Publ. 260-40 (August 1972.) COM72-50776**

Wood, L. A., and Shouse, P. J., Standard Reference Materials: Use of Standard Light-Sensitive Paper for Calibrating Carbon Arcs Used in Testing Textiles for Colorfastness to Light, NBS Spec. Publ. 260-41 (August 1972) COM72-50775**

Wagner, H. L. and Verdier, P. H., eds., Standard Reference Materials: The Characterization of Linear Polyethylene, SRM 1475, NBS Spec. Publ. 260-42 (September 1972). COM72-50944**

Yakowitz, H., Ruff, A. W., and Michaelis, R. E., Standard Reference Materials: Preparation and Homogeneity Characterization of an Austenitic Iron-Chromium-Nickel Alloy, NBS Spec. Publ. 260-43 (November 1972). COM73-50760**

Schooley, J. F., Soulen, R. J., Jr., and Evans, G. A., Jr., Standard Reference Materials: Preparation and Use of Superconductive Fixed Point Devices, SRM 767, NBS Spec. Publ. 260-44 (December 1972). COM73-50037**

Greifer, B., Maienthal, E. J. Rains, T. C., and Rasberry, S. D., Standard Reference Materials: Powdered Lead-Based Paint, SRM 1579, NBS Spec. Publ. 260-45 (March 1973). COM73-50226**

Hust, J. G., and Giarratano, P. J., Standard Reference Materials: Thermal Conductivity and Electrical Resistivity Standard Reference Materials: Austenitic Stainless Steel, SRM's 735 and 798, from 4 to 1200 k, NBS Spec. Publ. 260-46 (March 1975). SN003-003-01278-5

Hust, J. G., Standard Reference Materials: Electrical Resistivity of Electrolytic Iron, SRM 797, and Austenitic Stainless Steel, SRM 798, from 5 to 280 K, NBS Spec. Publ. 260-47 (February 1974). COM74-50176**

Mangum, B. W., and Wise, J. A., Standard Reference Materials: Description and Use of Precision Thermometers for the Clinical Laboratory, SRM 933 and SRM 934, NBS Spec. Publ. 260-48 (May 1974). 60 cents* SN003-003-01278-5

Carpenter, B. S., and Reimer, G. M., Standard Reference Materials Calibrated Glass Standards for Fission Track Use, NBS Spec. Publ. 260-49 (November 1974). COM74-51185

Hust, J. G., and Giarratano, P. J., Standard Reference Materials: Thermal Conductivity and Electrical Resistivity Standard Reference Materials: Electrolytic Iron, SRM's 734 and 797 from 4 to 1000 K, NBS Spec. Publ. 260-50 (June 1975). $1.00* SN003-003-01425-7

Mavrodineanu, R., and Baldwin, J. R., Standard Reference Materials: Glass Filters As a Standard Reference Material for Spectrophotometry; Selection; Preparation; Certification; Use-SRM 930, NBS Spec. Publ. 260-51 (November 1975). $1.90* SN003-003-01481-8

Hust, J. G., and Giarratano, P. J., Standard Reference Materials: Thermal Conductivity and Electrical Resistivity Standard Reference Materials 730 and 799, from 4 to 3000 K, NBS Spec. Publ. 260-52 (September 1975). $1.05* SN003-003-01464-8

Durst, R. A., Standard Reference Materials: Standardization of pH Measurements, NBS Spec. Publ. 260-53 (December 1975, Revised). $1.05 SN003-003-01551-2

Burke, R. W., and Mavrodineanu, R. Standard Reference Materials: Certification and Use of Acidic Potassium Dichromate Solutions as an Ultraviolet Absorbance Standard, NBS Spec. Publ. 260-54 (August 1977). $3.00* SN003-003-01828-7

Ditmars, D. A., Cezairliyan, A., Ishihara, S., and Douglas, T. B., Standard Reference Materials: Enthalpy and Heat Capacity; Molybdenum SRM 781, from 273 to 2800 K, NBS Spec. Publ. 260-55 (September 1977). $2.20* SN003-003-01836-8

Powell, R. L., Sparks, L. L., and Hust, J. G., Standard Reference Materials: Standard Thermocouple Materials, Pt.67: SRM 1967, NBS Spec. Publ. 260-56 (February 1978). $2.20* SN003-003-018864

Cali, J. P. and Plebanski, T., Guide to United States Reference Materials, NBS Spec. Publ. 260-57 (February 1978). $2.20* PB 277173

Barnes, J. D., and Martin, G. M., Standard Reference Materials: Polyester Film for Oxygen Gas Transmission Measurements SRM 1470, NBS Spec. Publ. 260-58 (June 1979) $2.00* SN003-003-02077

Chang, T., and Kahn, A. H. Standard Reference Materials: Electron Paramagnetic Resonance Intensity Standard; SRM 2601, NBS Spec. Publ. 260-59 (August 1978) $2.30* SN003-003-01975-5

Velapoldi, R. A., Paule, R. C., Schaffer, R., Mandel, J., and Moody, J. R., Standard Reference Materials: A Reference Method for the Determination of Sodium in Serum, NBS Spec. Publ. 260-60 (August 1978). $3.00* SN003-003 01978-0

Verdier, P. H., and Wagner. H. L., Standard Reference Materials: The Characterization of Linear Polyethylene (SRM 1482, 1483, 1484), NBS Spec. Publ. 260-61 (December 1978). $1.70* SN003-003-02006-1

Soulen, R. J., and Dove, R. B., Standard Reference Materials: Temperature Reference Standard for Use Below 0.5 K (SRM 768). NBS Spec. Publ. 260-62 (April 1979). $2.30* SN003-003-02047-8

Velapoldi, R. A., Paule, R. C., Schaffer, R. Mandel, J., Machlan, L. A., and Gramlich, J. W., Standard Reference Materials: A Reference Method for the Determination of Potassium in Serum. NBS Spec. Publ. 260-63 (May 1979). $3.75* SN003-003-02068

Velapoldi, R. A., and Mielenz, K. D., Standard Reference Materials: A Fluorescence Standard Reference Material Quinine Sulfate Dihydrate (SRM 936), NBS Spec. Publ. 260-64 (January 1980). $4.25* SN003-003-02148-2

Marinenko, R. B., Heinrich, K. F. J., and Ruegg, F. C., Standard Reference Materials: Micro-Homogeneity Studies of NBS Standard Reference Materials, NBS Research Materials, and Other Related Samples. NBS Spec. Publ. 260-65 (September 1979). $3.50* SN003-003-02114-1

Venable, W. H., Jr., and Eckerle, K. L., Standard Reference Materials: Didymium Glass Filters for Calibrating the Wavelength Scale of Spectrophotometers (SRM 2009, 2010, 2013). NBS Spec. Publ. 260-66 (October 1979). $3.50* SN003-003-02127-0

Velapoldi, R. A., Paule, R. C., Schaffer, R., Mandel, J., Murphy, T. J., and Gramlich, J. W., Standard Reference Materials: A Reference Method for the Determination of Chloride in Serum, NBS Spec. Publ. 260-67 (November 1979). $3.75* SN003-003-02136-9

Mavrodineanu, R. and Baldwin, J. R., Standard Reference Materials: Metal-On-Quartz Filters as a Standard Reference Material for Spectrophotometry-SRM 2031, NBS Spec. Publ. 260-68 (April 1980). $4.25* SN003-003-02167-9

Velapoldi, R. A., Paule, R. C., Schaffer, R., Mandel, J., Machlan, L. A., Garner, E. L., and Rains, T. C., Standard Reference Materials: A Reference Method for the Determination of Lithium in Serum, NBS Spec. Publ. 260-69 (July) 1980). $4.25* SN003-003-02214-4

Marinenko, R. B., Biancaniello, F., Boyer, P. A., Ruff, A. W., DeRobertis, L., Standard Reference Materials: Preparation and Characterization of an Iron-Chromium-Nickel Alloy for Microanalysis, NBS Spec. Publ. 260-70 (May 1981). $2.50* SN003-003-02328-1

Seward, R. W., and Mavrodineanu, R., Standard Reference Materials: Summary of the Clinical Laboratory Standards Issued by the National Bureau of Standards, NBS Spec. Publ. 260-71 (November 1981). $6.50* SN003-003-02381-7

Reeder, D.J., Coxon, B., Enagonio, D., Christensen, R. G., Schaffer, R., Howell, B. F., Paule, R. C., Mandel, J., Standard Reference Materials: SRM 900, Antiepilepsy Drug Level Assay Standard, NBS Spec. Publ. 260-72 (June 1981). $4.25* SN003-003-02329-9

Interrante, C. G., and Hicho, G. E., Standard Reference Materials: A Standard Reference Material Containing Nominally Fifteen Percent Austenite (SRM 486), NBS Spec. Publ. 260-73 (January 1982). $2.75* SN003-003-02386-8

Marinenko, R. B., Standard Reference Materials: Preparation and Characterization of K-411 and K-414 Mineral Glasses for Microanalysis: SRM 470. NBS Spec. Publ. 260-74 (April 1982). $3.50 SN003-003-023-95-7

Weidner, V. R., Hsia, J. J., Standard Reference Materials: Preparation and Calibration of First Surface Aluminum Mirror Specular Reflectance Standards (SRM 2003a), NBS Spec. Publ. 260-75 (May 1982). $3.75 SN003-003-023-99-0

Hicho, G. E. and Eaton, E. E., Standard Reference Materials: A Standard Reference Material Containing Nominally Five Percent Austenite (SRM 485a), NBS Spec. Publ. 260-76 (August 1982). $3.50 SN003-003-024-33-3

Furukawa, G. T., Riddle, J. L., Bigge, W. G., and Pfieffer, E. R., Standard Reference Materials: Application of Some Metal SRM's as Thermometric Fixed Points, NBS Spec. Publ. 260-77 (August 1982). $6.00 SN003-003-024-34-1

Hicho, G. E. and Eaton, E. E., Standard Reference Materials: Standard Reference Material Containing Nominally Thirty Percent Austenite (SRM) 487), NBS Spec. Publ. 260-78 (September 1982). $3.75 SN003-003-024-35-0

Richmond, J. C., Hsia, J. J. Weidner, V. R., and Wilmering, D. B., Standard Reference Materials: Second Surface Mirror Standards of Specular Spectral Reflectance (SRM's 2023, 2024, 2025), NBS Spec. Publ. 260-79 (October 1982). $4.50 SN003-003-024-47-3

Schaffer, R., Mandel, J., Sun, T., Cohen, A., and Hertz, H. S., Standard Reference Materials: Evaluation by an ID/MS Method of the AACC Reference Method for Serum Glucose, NBS Spec. Publ. 260-80 (October 1982). $4.75 SN003-003-024-43-1

Burke, R. W., Mavrodineanu, R. (NBS retired), Standard Reference Materials: Accuracy in Analytical Spectrophotometry, NBS Spec. Publ. 260-81 (April 1983). $6.00 SN003-003-024-84-8

Weidner, V. R., Standard Reference Materials: White Opal Glass Diffuse Spectral Reflectance Standards for the Visible Spectrum (SRM's 2015 and 2016). NBS Spec. Publ. 260-82 (April 1983). $3.75 SN003-003-024-89-9

Bowers, G. N., Jr., Alvarez, R., Cali, J. P. (NBS retired), Eberhardt, K. R., Reeder, D. J., Schaffer, R., Uriano, G. A., Standard Reference Materials: The Measurement of the Catalytic (Activity) Concentration of Seven Enzymes in NBS Human Serum SRM 909, NBS Spec. Publ. 260-83 (June 1983). $4.50 SN003-003-024-99-6

Gills, T. E., Seward, R. W., Collins, R. J., and Webster, W. C., Standard Reference Materials: Sampling, Materials Handling, Processing, and Packaging of NBS Sulfur in Coal Standard Reference Materials, 2682, 2683, 2684, and 2685, NBS Spec. Publ. 260-84 (August 1983). $4.50 SN003-003-025-20-8

Swyt, D. A., Standard Reference Materials: A Look at Techniques for the Dimensional Calibration of Sta..dard Microscopic Particles, NBS Spec. Publ. 260-85 (September 1983). $5.50 SN003-003-025-21-6

Hicho, G. E. and Eaton, E. E., Standard Reference Materials: A Standard Reference Material Containing Two and One-Half Percent Austenite, SRM 488, NBS Spec. Publ. 260-86 (December 1983). $1.75* SN003-003-025-41-1

Mangum, B. W., Standard Reference Materials: SRM 1969: Rubidium Triple-Point - A Temperature Reference Standard Near 39.30 °C, NBS Spec. Publ. 260-87 (December 1983). $2.25* SN003-003-025-44-5

Gladney, E. S., Burns, C. E., Perrin, D. R., Roelandts, I., and Gills, T. E., 1982 Compilation of Elemental Concentration Data for NBS Biological, Geological, and Environmental Standard Reference Materials. NBS Spec. Publ. 260-88 (March 1984). SN003-003-02565-8

Hust, J. G., A Fine-Grained, Isotropic Graphite for Use as NBS Thermophysical Property RM's from 5 to 2500 K, NBS Spec. Publ. 260-89 (In Press).

Hust, J. G., and Lankford, A. B., Update of Thermal Conductivity and Electrical Resistivity of Electrolytic Iron, Tungsten, and Stainless Steel, NBS Spec. Publ. 260-90 (In Press).

Goodrich, L. F., Vecchia, D. F., Pittman, E. S., Ekin, J. W., and Clark, A. F., Critical Current Measurements on an NbTi Superconducting Wire Standard Reference Material, NBS Spec. Publ. 260-91 (In Press).

* Send order with remittance to Superintendent of Documents, US Government Printing Office Washington, DC 20402. Remittance from foreign countries should include an additional one-fourth of the purchase price for postage.

** May be ordered from: National Technical Information Services (NTIS). Springfield Virginia 22151.

TABLE OF CONTENTS

		Page
Abstract		xiii
Disclaimer		xiii
1.	Introduction	1
2.	Experiment	2
	2.1 Sample Preparation	2
	2.2 Measurement Technique	3
	a. Temperature dependence	4
	b. Pressure effects	4
	c. Voltage-Current curve	5
	2.3 Raw Data Acquisition and Analysis	6
	a. Current ramp rate	7
	b. Other acquisition variables	8
	c. Raw data analysis	8
3.	Calibrations and Corrections	9
	3.1 Current	10
	3.2 Electric Field	10
	3.3 Magnetic Field	11
	3.4 Temperature	11
	3.5 Magnetic Field Profile	12
	3.6 Magnetic Field Angle	12
	3.7 Tensile Strain	12
	3.8 Bending Strain	13
	3.9 Time	16
4.	Preliminary Sample Screening	17
	4.1 Homogeneity and Other Key Properties	17
	4.2 SRM spools	19
5.	Sample Specimens and Measurements	19
	5.1 Range of Certified Values	19
	5.2 Measurement Procedure	20
	5.3 Variation with Distance	20
	5.4 Variation with Magnetic Field	23
	5.5 Variation with Temperature	25
	5.6 Variation with Electric Field	25
	5.7 Dependence on Electric Field and Temperature	25
	5.8 Dependence on Magnetic Field and Temperature	31
	5.9 Dependence on Magnetic Field and Electric Field	31
6.	Assumptions and Statistical Models	31
	6.1 Description of General Statistical Model	35
	6.2 Approximation of Temperature Dependence	35
	6.3 Statistical Model for Sample Data	36
7.	Statistical Analysis and Certification	36
	7.1 Material Inhomogeneity	36
	7.2 Tolerance Limits for Critical Current	37

TABLE OF CONTENTS (continued)

Page

7.3 Certified Critical Current at 4.2 K and 0.2 µV/cm 37

7.4 Temperature and Electric Field Corrections 38

7.5 Critical Temperature and Exponents 41

7.6 Additional Data Excluded from Certification 42

8. Discussion 43

9. Summary 46

10. References 48

Acknowledgments 48

Appendices 49

LIST OF FIGURES

Figure No.

Page

1   A schematic of the electrical systems for the I_c measurement   3

2   Typical dependence of I_c on ramp rate and voltage filter setting (2.55 and
     5.00 Hz). For a given symbol, the higher I_c is the up value (open symbol) . . .   7

3   Typical dependence of I_c on the number of points in the curve fit   9

4   Uniaxial strain degradation of I_c for three SRM specimens at 7 T. The
     shaded region is the range of I_c degradation for a number of NbTi conductors . .   14

5   Uniaxial strain degradation of I_c for a SRM specimen at different magnetic fields . .   15

6   The percentage change in I_c between the 5 and 50 m specimens of each
     sample, long-range homogeneity   18

7   Short-range homogeneity of the 5 and 50 m specimens of each sample. The
     50 m data are shifted slightly to the right and the 5 m data to the
     left over the sample number. Plotted are the percentage differences
     between the measured I_c on the center voltage tap and that of the other
     four taps (at 4 T)   18

8a   Critical current measurements at 2 T, 4.07 K, and 0.2 µV/cm: I_c (+) and
     average I_c (□) versus distance along wire   21

8b   Critical current measurements at 4 T, 4.07 K, and 0.2 µV/cm: I_c (+) and
     average I_c (□) versus distance along wire   21

8c   Critical current measurements at 6 T, 4.07 K, and 0.2 µV/cm: I_c (+) and
     average I_c (□) versus distance along wire   22

Page

8d   Critical current measurements at 8 T, 4.07 K, and 0.2 μV/cm: I_c (+) and average I_c ($\square$) versus distance along wire   22

9a   Critical current measurements at 3.90 K and 0.2 μV/cm: (I_c/average I_c) versus magnetic field   23

9b   Critical current measurements at 4.07 K and 0.2 μV/cm: (I_c/average I_c) versus magnetic field   24

9c   Critical current measurements at 4.24 K and 0.2 μV/cm: (I_c/average I_c) versus magnetic field   24

10a   Critical current measurements at 2 T and 0.2 μV/cm: I_c (+) and average I_c ($\square$) versus temperature   25

10b   Critical current measurements at 4 T and 0.2 μV/cm: I_c (+) and average I_c ($\square$) versus temperature   26

10c   Critical current measurements at 6 T and 0.2 μV/cm: I_c (+) and average I_c ($\square$) versus temperature   26

10d   Critical current measurements at 8 T and 0.2 μV/cm: I_c (+) and average I_c ($\square$) versus temperature   27

11a   Critical current measurements at 2 T and 4.07 K; natural log I_c (+) and average natural log I_c ($\square$) versus natural log of electric field criterion . . .   27

11b   Critical current measurements at 4 T and 4.07 K; natural log I_c (+) and average natural log I_c ($\square$) versus natural log of electric field criterion . . .   28

11c   Critical current measurements at 6 T and 4.07 K; natural log I_c (+) and average natural log I_c ($\square$) versus natural log of electric field criterion . . .   28

11d   Critical current measurements at 8 T and 4.07 K; natural log I_c (+) and average natural log I_c ($\square$) versus natural log of electric field criterion . . .   29

12a   Dependence of I_c on temperature and electric field at 2 T; points plotted are sample means of the natural logarithm of I_c   29

12b   Dependence of I_c on temperature and electric field at 4 T; points plotted are sample means of the natural logarithm of I_c   30

12c   Dependence of I_c on temperature and electric field at 6 T; points plotted are sample means of the natural logarithm of I_c   30

LIST OF FIGURES (continued)

Page

12d   Dependence of I_c on temperature and electric field at 8 T; points plotted
      are sample means of the natural logarithm of I_c   31

13a   Dependence of I_c on temperature and magnetic field at 0.05 µV/cm; points
      plotted are sample means   32

13b   Dependence of I_c on temperature and magnetic field at 0.10 µV/cm; points
      plotted are sample means   32

13c   Dependence of I_c on temperature and magnetic field at 0.20 µV/cm; points
      plotted are sample means   33

14a   Dependence of I_c on electric field and magnetic field at 3.90 K; points
      plotted are sample means   33

14b   Dependence of I_c on electric field and magnetic field at 4.07 K; points
      plotted are sample means   34

14c   Dependence of I_c on electric field and magnetic field at 4.24 K; points
      plotted are sample means   34

15a   Dependence of I_c on electric fields from 0.02 to 0.5 µV/cm at 2 T; natural
      logarithm of I_c and electric field are plotted   44

15b   Dependence of I_c on electric fields from 0.02 to 0.5 µV/cm at 4 T; natural
      logarithm of I_c and electric field are plotted   45

15c   Dependence of I_c on electric fields from 0.02 to 0.5 µV/cm at 6 T; natural
      logarithm of I_c and electric field are plotted   45

15d   Dependence of I_c on electric fields from 0.02 to 0.5 µV/cm at 8 T; natural
      logarithm of I_c and electric field are plotted   46

LIST OF TABLES

Table No.   Page

1   Estimated limits to systematic and Random Errors Expressed as Percent Error
    in I_c at Magnetic Fields of 2, 4, 6, and 8 T   10

2   Estimated Long- and Short-Range Inhomogeneity Expressed in Percent Error in I_c . . .   37

3   Critical Current at 4.2 K and 0.2 µV/cm   38

Table No.   Page

4   Coefficients for Temperatures and Electric Field Corrections   38

5   Critical Current at 0.05, 0.10, 0.20 µV/cm and 3.90, 3.91, ..., 4.24 K   39

6   Critical Temperature (T_c*) and Exponent (n)   42

7   Critical Currents at 0.02 µV/cm   43

8   Critical Currents at 0.5 µV/cm   44

ABSTRACT

This report reviews the selection and certification by NBS of a Standard Reference Material (SRM) for the measurement of superconducting critical current. Procedures for preparing and measuring five candidate conductors are described. Evaluation criteria are discussed by which one of the five conductors was selected for the critical current SRM. The designated superconducting wire, SRM 1457, has been subdivided and wound onto 500 spools for distribution. Certified critical current measurements were made on a sample of these spools. Material variability, or inhomogeneity, along the whole wire is included in a statistical model based on the dependence of critical current on temperature and electric field. Critical currents for SRM 1457 are certified at magnetic fields of 2, 4, 6, and 8 T for temperatures from 3.90 to 4.24 K and electric field criteria from 0.05 to 0.2 μV/cm. Statistical tolerance limits and estimated systematic errors are combined to give an overall uncertainty in the certified values. The total uncertainty is no greater than 2.57 percent of the reported critical current at any of the four magnetic fields.

Key words: Critical current; electric field criteria; inhomogeneity; magnetic field; NbTi; standard reference material; statistical model; statistical tolerance interval; superconductor; temperature; uncertainty.

DISCLAIMER

An effort was made to avoid the identification of commercial products by the manufacturer's name or label, but in some cases these products might be indirectly identified by their particular properties. In no instance does this identification imply endorsement by the National Bureau of Standards, nor does it imply that the particular products or equipment are necessarily the best available for that purpose.

The criteria used to select the conductor to be the SRM were developed only for this application. This conductor might not be the best for any other application and, in fact, it might not be the best for this application, due to the limitation of the preliminary screening. The possible problems which eliminated candidate conductors might not be present throughout the conductors or they might not adversely affect other applications.

CRITICAL CURRENT MEASUREMENTS ON AN NbTi SUPERCONDUCTING WIRE STANDARD REFERENCE MATERIAL*

L. F. Goodrich, D. F. Vecchia, E. S. Pittman, J. W. Ekin, and A. F. Clark

1. INTRODUCTION

The National Bureau of Standards is producing a standard reference material (SRM) for the measurement of the superconducting critical current (I_c). This SRM and the recently adopted American Society for Testing and Materials (ASTM) critical current standard test method [1][+] will aid both the commerce and technology of superconductors through the promotion of more uniform measurements. The SRM will serve as a means for interlaboratory comparison that will further advance the consensus and evolution of the new test method. The general use and philosophy of an SRM are given elsewhere [2].

To perform well as an SRM, the conductor chosen should be as homogeneous a composite as possible. Filamentary niobium-titanium (NbTi) superconductors with a copper matrix were purchased from the inventories of each of the United States wire manufacturers. Each was selected by the manufacturer as a good candidate for an SRM. Each candidate conductor was provided in a continuous length. NbTi was chosen rather than niobium-tin because the I_c of the NbTi is much less strain dependent and the conductor homogeneity is considered to be better. Preliminary screening measurements were performed on each conductor to determine the short- and long-range spatial variations in the critical current. The choice of the SRM was based primarily on these data.

The conductor selected for the critical current SRM was wound onto 500 distribution spools, each with ~2.2 m of wire. Measurements on a sample of the spools were used to determine the likely outcome of measuring the I_c of any random spool. The uncertainty in certified values applies when the user makes one I_c measurement, on one spool, at one of the given magnetic fields (2, 4, 6, and 8 T), at any temperature from 3.90 to 4.24 K, and using any electric field criterion from 0.05 to 0.2 µV/cm.

A description of the I_c measurement procedure is presented. Sample handling and mounting are described. A number of acquisition and analysis variables were identified and their effect on the precision and accuracy of the critical current measurement determined. The chosen SRM was tested for uniformity and the effect of several measurement variables was determined. Data on the variations in the I_c along the length of the conductor are given.

---

*Work supported in part by the NBS Office of Standard Reference Materials and the Department of Energy.

[+]A copy of the separate standard test method can be obtained by writing to ASTM Customer Service, 1916 Race Street, Philadelphia, Pennsylvania 19103, or calling (215) 299-5585 [TWX: 710-670-1037].

2. EXPERIMENT

The only significant change from the common instrumentation [3] used to measure I_c was the addition of a digital processing oscilloscope. This allowed automation of the data analysis and achievement of higher precision in the processing of the voltage-current (V-I) curves than that achieved by using an analog X-Y recorder alone. A number of acquisition and analysis variables were identified and their effect on the precision and accuracy of the critical current measurement determined. A correction for the magnetic field profile was made to I_c measured on adjacent voltage taps. The hydrostatic head and stratification of the liquid helium bath were also measured and the correct liquid helium vapor pressure used for the temperature determination.

2.1 Sample Preparation

The principal sample geometry used was a helical coil with a pitch of about 2 turns per cm. The coil mandrel diameter was 3.18 cm, so the sample bending diameter was approximately 3.23 cm. The coil mandrel was made from a G-11 fiberglass-epoxy tube with a circumferential fiber direction (rolled tube) so that the specimen and holder had similar thermal contraction characteristics. This choice of sample geometry should not affect the value of I_c measured, according to a previous study [3], except for the bending strain effect. There are many tradeoffs for each geometry (e.g., long straight, short straight, hairpin, etc.), but the coil seemed the best for this type of testing. With the coil, a number of segments of a given specimen could be measured to test the short-range variations. Also the problems of negative voltage [4], magnetic field angle and field uniformity are fewer in this geometry. The only major problem is the effect of bending strain, which is not well known and is hard to measure for NbTi superconductors. A discussion of the effect of bending strain is given in Section 3.8. Because the bending strain correction is not known exactly, the correction of the I_c data to the unbent state was not made, but estimates of its effect are given.

The sample was prepared for mounting by unreeling a length just over one meter and taping it to a flat surface. A phenol/methylene chloride wire stripping compound was applied to the ends of the sample and to the locations of the voltage taps. The specimen holder was coated with a slow drying varnish. After removal of the insulation, one end was hooked to a tensional wire fixed to the sample holder. The sample wire was then coiled around the upper copper current contact of the mandrel and onto the epoxy-glass fixture that had a spiral groove machined into its surface to retain the wire in proper orientation. Tension was maintained on the wire by pulling it tangentially outward as it was threaded into this groove. At the other end of the epoxy-glass mandrel the wire was coiled around the lower copper current contact and hooked to another tensional wire fixed at this end. Sufficient tension was maintained with these wires so that the specimen does not move while the underlying varnish was drying. After the varnish was dry, solder was flowed liberally on both copper current contacts, wetting the ends of the specimen wire (not previously tinned) and fixing it permanently in place. Voltage taps, spaced approximately 2 cm apart, were then soldered in place, and the spacing measured with a ruler. All connecting tap wires are then taped to the mandrel. Finally, with one ampere flowing through the specimen, the voltage drop across each set of taps was measured and logged. This was used to confirm the lack of anomalies in tap placement and connection. The solder used for the current contacts and the voltage taps was a tin-lead eutectic alloy (Sn63%, Pb37%) with resin flux (meets Federal Specification QQ-S-571). Other tin-lead solders (Sn-60Pb and Sn-50Pb) have been used in other experiments without detrimental effects.

2.2 Measurement Technique

A description of some of the details of the instrumentation follows here. A more complete
description of the measurement parameters is given elsewhere [3]. A block diagram of the electrical
system is shown in figure 1. The magnet is a superconducting, 9 T, 3.8 cm bore solenoid. The sample
current battery power supply, standard resistor and vapor-cooled leads have the capacity to deliver
600 A into the sample. The quench detector can reset the ramp generator, and thus shut down the
sample current, within 1 ms of detection of the normal state (quench). A chopper-type, analog
nanovoltmeter with an amplified analog output was used as the first stage in the sample voltage
amplifier circuit. The active, low-pass filter was used to attenuate the low frequency beating of the
harmonics of the chopper frequency and the line frequency noise. An X-Y recorder was used to monitor
the voltage-current (V-I) curves and to create a hard copy of the data. The voltage and current
signals were further amplified and fed into simultaneous channels of a digital processing oscilloscope.
The critical current was determined from these V-I curves. The digital processing oscilloscope
allowed automation of the data analysis and higher precision processing than an X-Y recorder alone.
Pressure in the helium dewar was measured with an absolute pressure gauge (Bourdon tube type). The
pressure control was obtained with a Cartesian manostat. A vacuum pump was connected to the manostat
output for operation below atmospheric pressure.

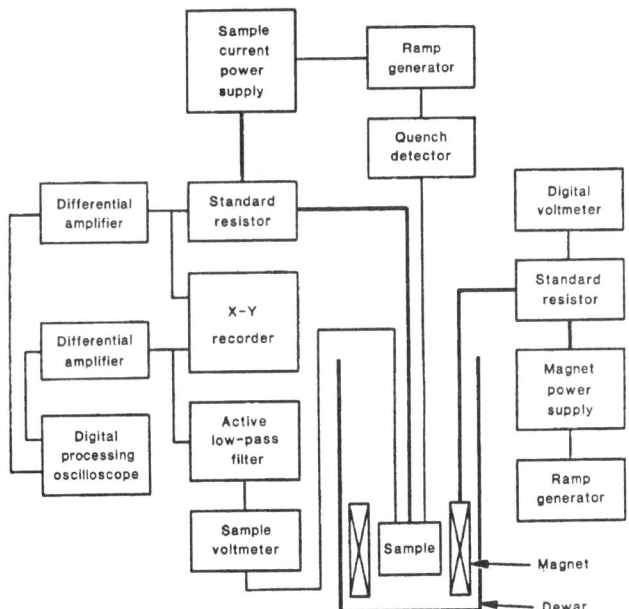

Figure 1. A schematic of the electrical systems for the I_c measurement.

a. Temperature dependence

The temperature dependence of I_c is included in the statistical model to allow the user to measure I_c at any temperature, T, (helium vapor pressure) within the given range. The approximate empirical relationship at constant magnetic field is linear [3],

$$I_c = mT + b. \tag{1}$$

This equation becomes

$$I_c = m(T - T_c^*), \tag{2}$$

with T_c^* defined as the temperature at which I_c is zero (the critical temperature for a given magnetic field). Using the expression at a reference temperature, T_r, and an associated I_r, to substitute for m, the equation becomes

$$I_c = I_r [1 + \frac{T_r - T}{T_c^* - T_r}]. \tag{3}$$

This equation is used later in the statistical analysis.

The critical current was measured at three temperatures: 3.90, 4.07, and 4.24 K. The values used for the statistical analysis were: 3.8990, 4.06995, and 4.24273 K; however, in the text they are rounded to three significant digits. This range of temperatures was chosen to cover the expected range of operating temperatures. Most of the measurements were made at the center temperature. The temperature dependence was determined using all of the measurements. Because of the high currents involved, the measurement is best conducted in a constant temperature bath of liquid helium. The easiest way to keep a bath at constant temperature is with a pressure controlling manostat. A thermometer could be used to determine the temperature of each I_c measurement, but the complication of magnetic field effects on the thermometer calibration makes this undesirable. So the sample temperature was determined by measuring the helium vapor pressure and using a convenient temperature scale, the 1958 ^{4}He vapor pressure scale. A comparison of this temperature scale with others is given elsewhere [5,6].

b. Pressure effects

There is an experimental difficulty in making measurements above atmospheric pressure because of the natural stratification of the liquid helium. This results when warmer, less dense, higher vapor pressure liquid helium is on top of cooler, more dense, lower vapor pressure liquid. The higher two temperatures, 4.07 and 4.24 K, required pressurizing above ambient at the testing site. It would take several hours for conductive and convective mixing to return the helium bath to thermal equilibrium from a stratified state. A heater on the bottom of the helium dewar was used to warm the liquid and bring the whole bath to equilibrium at a pressure about 1333 Pa (10 mmHg) above the desired pressure, at which point the heater was turned off. After a settling time of a few minutes, the controlled pressure was reduced to the desired pressure. The liquid temperature was monitored with three carbon resistance thermometers at various levels in the liquid to observe the response to these pressure changes. The bath responds without stratification as the pressure is reduced, but up and down variations in the pressure can still result in uncertainties in the temperature. Variations in pressure occur because ramping the magnet and sample current changes the liquid helium boil-off rate causing the dewar ullage pressure to change slightly. Typical pressure changes were less than 267 Pa (2 mmHg) with a temperature change at the sample of less than 3 mK.

Other effects in a liquid helium bath are the hydrostatic head effect [7] and the possible magnetic-field-induced hydrostatic head effect [8]. The latter, if it is present, is negligible in this experiment. The hydrostatic head of the liquid helium column was observed by measuring the temperature as a function of depth into the liquid with a carbon thermometer on the end of an adjustable depth probe. This experiment also verified that the technique used to counteract the stratification worked. A temperature gradient was observed only near the surface of the liquid and it was smaller than calculated, probably because of conductive and convective mixing [7]. For a helium depth of about 59 cm, the total temperature change was only 2.3 mK, whereas the calculated value was 8.8 mK. The temperature gradient near the surface appeared to be reduced when the liquid level was close to the top of the superconducting magnet. This was probably due to the increased conduction and convection in that region of the dewar. One of the carbon resistors was located out of the high magnetic field region at the bottom of the dewar so that the magnetic field had little effect on it. This resistor was monitored throughout each data run and these measurements confirmed the observations of the temperature profile experiment.

c. Voltage—Current curve

Voltage versus current data were generated for each magnetic field (2, 4, 6, and 8 T) and for each pair of voltage taps. In practice, voltage versus time and current versus time curves were recorded. The voltage tap leads were continuous from the sample to the input cord of the nanovoltmeter. The junction between the voltage tap leads and the nanovoltmeter input leads was maintained isothermal by putting small bags of lead shot under and on top of the junction. The amplified signal (gain of 100,000) was then passed through an active low-pass filter and an instrument amplifier which provided an additional gain of 10 as well as being balanced to single-ended conversion to match the input of the digital processing and storage oscilloscope. The output from the 1000 A, 100 mV current-sensing resistor (current shunt) was connected directly to an instrument amplifier with a gain of 100 that was then fed into the other channel of the oscilloscope. The oscilloscope trace was manually triggered and began to store 2048 data points at 100 ms per point (data frame of 204.7 s). The scope was allowed to run for about 30 seconds while the sample current was zero in order to acquire baseline information. After switching from one pair of voltage taps to another, the voltage baseline was monitored until the thermoelectric voltage reach equilibrium. The current supply was then ramped up at a rate sufficient to reach about 80% of the critical current in about 30 seconds. At this point the ramp rate was reduced by a factor of 10. The ramp continued until just short of the thermal runaway current at which point the ramp was reversed. The current continued to ramp downwards until the current supply output reached zero. The scope was then allowed to acquire baseline data for another 30 seconds. Data acquisition was then halted and field, voltage tap, other information and both 2048 point digital records of current and voltage were written to a floppy disk for later analysis.

The dependence of I_c on electric field criterion, E_c, is included in the statistical model to allow the user to measure I_c at any criterion within a given range. It was also used in the analysis of the V-I curves. The electric field was defined as the measured voltage divided by the separation of the voltage tap pair. The approximate empirical relationship at constant magnetic field is

$$E_c = a \, I_c^{\,n},\qquad(4)$$

where a is a proportionality constant and the power n is typically 40 to 60, depending on the magnetic field and the sample [3]. Using a reference electric field, E_r, and an associated I_r, the equation becomes

$$I_c = I_r \, (\frac{E_c}{E_r})^{1/n}.\qquad(5)$$

This equation is used later in the statistical analysis.

The stored digital records of current and voltage were analyzed to determine the critical current at various criteria for both increasing and decreasing current. The current channel data were scaled in amperes and the voltage channel data was changed to electric field and scaled in microvolts/cm. The average value of the initial baseline of both records was determined and used to zero any constant amplifier offset and thermoelectric voltage. The change in the baseline of the voltage record was assumed to be a linear drift and this first order correction was later added to the level of electric field criteria. A typical value for the thermoelectric correction was 2 nV/cm. In order to make a first order correction to the electric field induced on the voltage taps due to the time rate of change of current (dI/dt correction), the electric field and the dI/dt were measured for the fast ramp region of the waveforms and the proportionality constant determined. This assumed that the electric field in this low current region was only that induced, which was a good assumption for this geometry. With this proportionality constant and the dI/dt in the critical current region of the waveform, a dI/dt correction to the level of the electric field criteria could be made. A typical value for the dI/dt correction was 2 nV/cm. The section of the electric field waveform around I_c was then separated and the logarithm of the electric field was taken. The logarithm changes the shape of the electric field versus time from a curve, to more of a linear relationship. This was due to the approximate power law relationship between electric field and current, and the approximately constant dI/dt. With the approximately linear waveforms, straight line fitting of segments could be performed, which can smooth the noise and interpolate between the digital points of the waveform. To determine I_c using the digitized waveforms, the point nearest to a given electric field level, with corrections, was found and a certain number of points on each side of this one were included in a linear fit. From this fit, the time at which the electric field was equal to the set point was determined. Then by fitting the simultaneous section of the current waveform, the corresponding current was determined. These steps were performed for the chosen set of electric field criteria on both the increasing and decreasing current portions of the waveforms.

2.3 Raw Data Acquisition and Analysis

A number of acquisition and analysis variables were identified and their effect on the precision and accuracy of the critical current measurement determined. The acquisition variables considered were: current ramp rate, digital sampling rate, and voltage filtering and amplification. The analysis variables considered were: corrections for the inductive voltage and changes in thermoelectric voltage, and type and extent of curve fitting. Some of these variables were mentioned above but will be discussed again in this context.

a. Current ramp rate

One important variable is the current ramp rate. It effects the critical current determination and many other acquisition and analysis variables. The I_c of interest is the steady state value but, in practice, it is determined with a relatively slow ramp. There are tradeoffs on both ends of the ramp rate. For the fast rates the response of the voltmeter and filtering limit the accuracy of I_c. At the slow end, the time necessary for the measurement limits the number of measurements that can be made in a day. A combination of a fast rate for the low current and a slow ramp near I_c was chosen, as it combines some of the advantages of each. For the low current, the voltage was not of much interest since it was mostly due to dI/dt for this geometry and was fairly constant with current. At the ramp rate change, at approximately 80% of I_c, the voltage returned to very close to zero; it had only a small offset due to the dI/dt. With this slow finishing ramp rate, many points were digitized in the region of I_c. A typical final rate was equivalent to $I_c/150$ A/s and the fast rate was ten times that.

The effect of current ramp rate on the I_c determination was studied. Two questions were addressed, what limitation does the equipment have and was there any fundamental dependence of I_c on the ramp rate for a reasonable range of ramp rates. This was studied by measuring I_c, for both the increasing

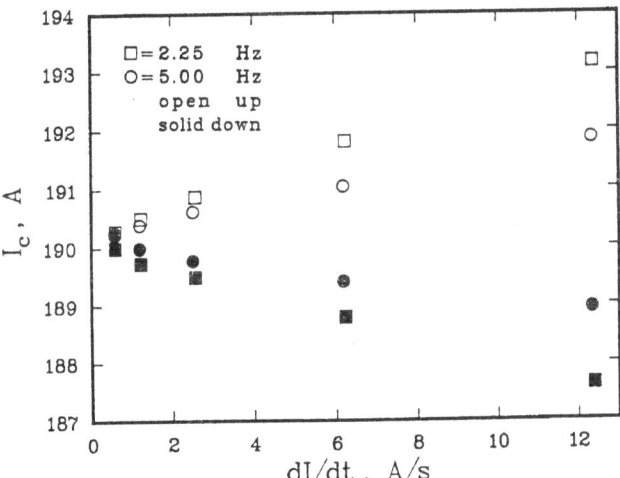

Figure 2. Typical dependence of I_c on ramp rate and voltage filter setting (2.55 and 5.00 Hz). For a given symbol, the higher I_c is the up value (open symbol).

and decreasing current portions (up and down I_c) of the V-I curve, as a function of ramp rate and active filter setting. Typical data are presented on figure 2. The "up" I_c is indicated by an open symbol and the "down" I_c, a lower value, by a solid symbol. Similar data were obtained for all five candidate conductors, at other magnetic fields, and at other electric field criteria. Each point on this figure is the average of five observations. The range of the observations for each point was less than 0.2%. The normal rate for this field was 1.25 A/s or zero to I_c in 152 s. The approximate range of ramp rates was from one-half to ten times the normal rate. A reasonable method of removing the effect of the filtering and response of the voltmeter is to average the up and down I_c values which gives the same I_c for all rates to within ±0.1%. Therefore, if I_c has a dependence on ramp rate it is less than 0.1% for this range of ramp rates once the filtering and response time have been removed. For all of the magnetic fields, the finishing ramp rate was equivalent to, from zero to I_c in about 150 s. The suggested limits on the ramp rate for the SRM are from one-half to five times this rate with the filtering and response time correction.

b. Other acquisition variables

The rest of the acquisition variables were more easily determined. The normal digital sample rate was 100 ms per point, which gave a comfortable window for each observation. For the ramp rate dependence study, the sampling rate was changed with the same 1, 2, 5 sequence as the ramp rate. The voltage filtering used was the 2.25 Hz setting of the active filter which gave smooth curves that were easy to find voltage levels on. The choice of gain on the voltage and current channels was based on the residual noise of the differential amplifiers. The analog to digital converters were biased so that the maximum dynamic range of 12 binary bits (4096) was used.

c. Raw data analysis

In the analysis of the digital V-I curves, the effects of the corrections and variables on the precision and accuracy of the I_c determination were considered. The typical size of both the inductive voltage and the thermoelectric voltage corrections to the electric field criteria, was 2 nV/cm. These corrections were relatively small and thus had little effect on the precision and accuracy, except in the ramp rate study where the inductive voltage corrections were significant. In that case, the correction clearly improved the accuracy of the measurement. Because the logarithm of E was not perfectly linear with time, there was concern as to when increasing the number of points in the linear fit would start to increase the precision at the expense of accuracy. In order to study this, a set of five repeat observations on the same tap, in the same field, were analyzed with a different number of points in the fit. Typical results are shown on figure 3. Similar results were obtained at other electric field criteria, other ramp rates and other magnetic fields. Notice that as the number of points increases, the range of scatter in the data decreases to an acceptable level of precision. At the same time, the average value does not change significantly with the number of points, indicating that accuracy is not sacrificed to gain this precision. The 35 point fit (17 points on each side of the electric field level) was used in all of the rest of the analysis. A disadvantage of extending the fit more is that at some point below the lowest electric field criterion, the voltage can get very close to zero. This changes the scale of the logarithm of E and the dynamic range of the digital waveform is not efficiently used. Also, it starts to limit the highest electric field criterion that can be analyzed because the current is conservatively turned around before thermal runaway occurs. Only certain numbers of points in the fit were considered because of the presence of the small amplitude,

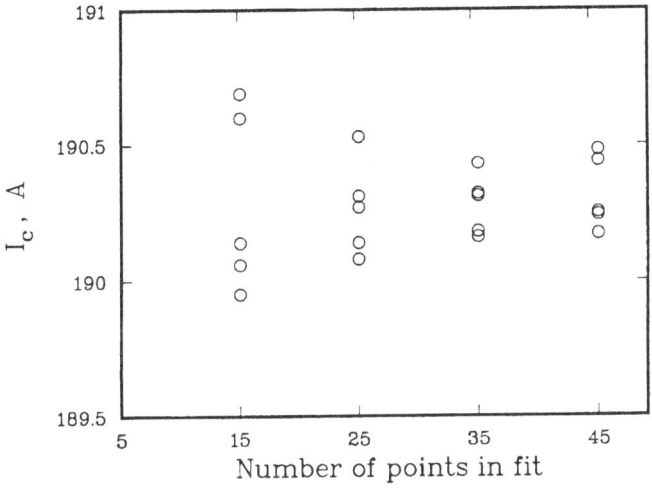

Figure 3. Typical dependence of I_c on the number of points in the curve fit.

low frequency beating of the nanovoltmeter chopper and the line frequency noise which is strobed by the digital sampling rate. These numbers were determined to keep close to an integer multiple of this beat frequency in the fit and to keep it centered around the electric field level.

3. CALIBRATIONS AND CORRECTIONS

The calibration of the instruments used to measure voltage, current, magnetic field, pressure, and length was performed. The estimated limits to systematic errors (inaccuracies) and random errors (precision limits) are given in table 1. The random errors are included in the measured variation in the critical current and for this reason were separated from the systematic errors. The error in each of the critical current variables is expressed in terms of the resultant percentage error in the critical current at each of the magnetic fields. These errors were estimated using the known dependence of the critical current on each of the variables. The periodic and random deviation (PARD) of the current and magnetic field (called out in the ASTM specification) are not included because the sample current source was a battery power supply and the magnet was used in the persistent mode, so these are both zero in this experiment. The sum and the root-mean-square are given at the bottom of the table for each magnetic field. The components of error for each variable were added and the root-mean-square of the variable errors was used for the total systematic error.

The philosophy followed was to correct to an ideal state, rather than to give the results under the arbitrary testing conditions. The only exception to this was the bending strain of the samples. That correction was not made because of the large uncertainty in the effect. The correction of the response time of the voltmeter, hydrostatic head, etc., caused a number of small corrections to the I_c

data. Many of these small effects are present in any measurement, but would not be exactly the same
as they are in this case. The corrections considered are listed below and the uncertainty of each
correction was included in the estimated systematic error of the relevant variable.

3.1 Current

The systematic and random errors of the current shunt and the shunt voltage input channel (signal
conditioner, amplifier and analog-to-digital converter) are directly reflected as errors in I_c. The
gain of the input channel was checked four times throughout the course of the I_c measurements with a
calibrated voltage source. The total systematic error in I was taken as the sum of the estimated
calibration uncertainties of the shunt (including the temperature dependence), the input channel gain,
and the voltage source. The random error was due to the sum of the input channel noise, variations in
shunt temperature, and thermoelectric voltages.

3.2 Electric Field

Errors in the electric field are relatively large because the voltages are small and the separation of the voltage taps is hard to measure, but I_c is relatively insensitive to errors in E. For
example, a 5% change in E would only change I_c by about 0.12% for a magnetic field of 8 T. The errors
in E will affect I_c by different amounts at different magnetic fields because of the shape changes in
the V-I curve. The voltage input channel was checked four times throughout the course of the I_c
measurements with a calibrated voltage source, just as the shunt voltage input channel was. As stated
earlier, the inductive and thermoelectric voltage corrections were small, typically 2 nV/cm, so the
systematic and random error of these corrections are negligible. The I_c measured with increasing
current was averaged with I_c measured with decreasing current at a given electric field criterion in
order to correct for the response time of the voltmeter and filter. The estimated inaccuracy of this
correction was included in the systematic error of E. Other systematic errors were the estimated

Table 1. Estimated limits to systematic and Random Errors Expressed as Percent
Error in I_c at Magnetic Fields of 2, 4, 6, and 8 T.

Variable	Systematic error (%)				Random error (%)			
	2 T	4 T	6 T	8 T	2 T	4 T	6 T	8 T
Current	0.10	0.10	0.10	0.10	0.05	0.05	0.05	0.05
Electric field	0.08	0.08	0.08	0.10	0.05	0.05	0.06	0.07
Magnetic field	0.17	0.20	0.34	0.74	0.04	0.05	0.05	0.08
Temperature	0.03	0.04	0.05	0.09	0.10	0.12	0.16	0.27
Magnetic field profile	0.02	0.03	0.05	0.10	0.01	0.01	0.01	0.01
Magnetic field angle	0.10	0.10	0.10	0.10	0.01	0.01	0.01	0.01
Tensile strain	0.04	0.05	0.06	0.08	0.01	0.01	0.02	0.02
$\Sigma\Delta$	0.54	0.60	0.78	1.31	0.27	0.30	0.36	0.51
$\sqrt{\Sigma(\Delta^2)}$	0.24	0.27	0.39	0.78	0.13	0.15	0.19	0.30

calibration uncertainties of the voltage input channel, the voltage source, and the ruler used to measure the voltage tap separation. The random error was due to the sum of the input channel noise, sample motion, and random error in the length measurement.

The range of electric field criteria obtained was restricted on the low end by noise, thermoelectric and inductive voltages; and on the high end by the reversal of the V-I curve and the requirements of the curve fitting routine. Under these conditions, the lowest criterion was 0.05 µV/cm that was measured in both increasing and decreasing current. The value of I_c at 0.02 µV/cm was measured for increasing current only because the inductive voltage would occasionally limit the fitting routine. The current was conservatively reversed before thermal runaway occurred to reduce the flux-flow heating in the sample and to avoid occasional quenches. This created censored holes in the data set at 0.5 µV/cm, especially at the higher magnetic fields. So the certified range of electrical field was from 0.05 to 0.2 µV/cm with informational values at 0.02 and 0.5 µV/cm. This is consistent with the trend toward the use of criteria lower than the 1 µV/cm suggested for NbTi [1].

The voltage tap separation of approximately 2 cm was chosen because it was considered to be close to the shortest length that would be measured and, thus, would give a representative maximum variation in I_c. For lengths larger than 2 cm, the variation in I_c is expected to be less. For lengths shorter than 2 cm, the variation in I_c may be more.

3.3 Magnetic Field

The largest systematic error in I_c is due to uncertainty in calibration of the magnetic field. The magnetic field of the superconducting background magnet was measured with a rotating coil gaussmeter calibrated with a standard magnet. The temperature coefficient of the probe was small enough that the slight cooling of the probe in the dewar was insignificant. Along with the current-central field relationship of the magnet, the axial profile measurements and some off-axis measurements were made. The off-axis measurements were used to determine the calibration for the off-axis sample position. A small correction was made to the I_c data for the magnetic field created by the current in the sample coil itself. This effect was determined by measuring I_c for both sample current directions at each magnetic field and it was only significant at the lowest applied field, 2 T. The estimated systematic error of these corrections was included in the systematic error in the magnetic field variable. Other sources of systematic error were the standard magnet calibration, and the accuracy of the rotating coil gaussmeter (linearity). The constant percent uncertainty caused the systematic error to increase with magnetic field. The sources of random error were from setting the magnet current, thermoelectric voltage, and a very slight decay of magnet current with time resulting from a finite resistance in the persistent mode.

3.4 Temperature

The largest random error was in the temperature variable. This was due to the combination of liquid helium stratification, pressure reading, and hydrostatic head variations. A correction was made to the pressure reading assigned to each I_c measurement to account for the hydrostatic head. This correction was based on the recorded value of the carbon resistor for each I_c measurement. The correction was always quite small, typically less than 3 mK and even less for the higher magnetic field measurements which were made at the end of the data run. The temperature of each I_c measurement

11

was not exactly at the desired value because of the hydrostatic head or changing control pressure, so the value of I_c was adjusted to this constant temperature using the known temperature dependence. This adjustment was small; its value at the highest field was typically only 0.1%. The systematic error of the temperature variable was due to the uncertainty in the hydrostatic head correction, pressure meter calibration, and the magnetic field induced hydrostatic head.

The thermal runaway (quench) of all five candidate conductors occurred at relatively low electric fields of about 1 to 2 μV/cm (the lower value at the higher magnetic fields). This fact limited the range of electric fields that could be measured. Sample motion can cause early or low runaway and can also falsely trip the quench detector. So, alternate sample mounting techniques were investigated. It was observed that coating the coil mandrel with petroleum jelly (petrolatum, which holds the sample at 4 K) or varnish, did not significantly change the thermal runaway level, nor did it change I_c. This indicated that sample motion was not a problem. Another experiment was performed on a number of specimens to determine if thermal runaway occurs or if the quench detector was giving tripping falsely. This was accomplished with fast acquisition of the sample voltage from three sections of the sample simultaneously using three of the four digital processing oscilloscope channels and using the fourth channel to monitor and trigger on the current signal. A thermal runaway voltage was observed to initiate in different places for each specimen, but it was not observed to be adjacent to a current contact. The slight heating at the current contacts (10 cm long) and the resulting temperature profile along the conductor will lower I_c at the sample ends relative to the central region. Apparently this effect is more than compensated for by the increase in I_c at the sample ends due to the magnetic field profile. A weak link will cause an apparent early thermal runaway if it is not in the region being measured. The length in high, perpendicular magnetic field (the active length), was about 1 m for the coil specimen holder used here. A small change in I_c results in a large change in the electric field. For example a 0.5% change in I_c changes the electric field by about 30%. For a sample geometry with a shorter active length, a higher electric field runaway might be observed because of a possible smaller variation in the local I_c.

3.5 Magnetic Field Profile

The profile of the magnetic field caused the measurements on different voltage taps to give slightly different values for I_c. The value of I_c was corrected to a single magnetic field using the measured profile and the magnetic field dependence of I_c. The uncertainty in the profile measurement was used to estimate the systematic error of this correction. Variation in the sample voltage tap placement and cryostat position accounted for the random error associated with this variable.

3.6 Magnetic Field Angle

The known dependence of I_c on the angle of the magnetic field was used to correct I_c to the orthogonal value. The systematic error of this variable was taken as the uncertainty of this correction. The possible variation in angle determined the random error.

3.7 Tensile Strain

The measured dependence of I_c on tensile strain was used to make a correction to I_c for the estimated winding tension of the sample mounting. The uncertainty of this correction and the uncertainty

12

in the differential thermal contraction account for the systematic error in I_c due to this variable. The estimated variation in winding tension determined the random error. The experiment was designed such that the Lorentz force was directed into the specimen holder thus eliminated hoop stress due to the applied magnetic field.

3.8 Bending Strain

Bending the sample into a coil can degrade the critical current of the conductor. The critical current results reported here were obtained for a bend diameter of 3.23 cm (coil mandrel diameter of 3.18 cm). If the SRM is measured with a bend diameter other than 3.23 cm, the results may be slightly different. For a bend diameter from infinity (straight) to about 2 cm, the certified I_c values can be used only if the total uncertainty is increased by an additional amount:

$$G\left|1 - (\frac{3.23}{d})^2\right|. \tag{6}$$

Here G is 1.1% at 2 T, 1.2% at 4 T, 1.3% at 6 T, and 1.5% at 8 T.

Equation (6) is obtained as follows. It has been shown experimentally that the bending strain degradation of I_c relative to its maximum (strain-free) value I_{cm} is given approximately by [9]:

$$(I_c/I_{cm})_{bending} = 1 - \frac{\alpha}{d^2}. \tag{7}$$

Here d is the diameter of the sample and the parameter α represents the sensitivity of I_c of the sample to bending strain. Thus, the difference between the bend effect at a bend diameter of 3.23 cm and an arbitrary bend diameter d is given by eq (6) with:

$$G \equiv \alpha(3.23)^{-2}. \tag{8}$$

The values of G listed above are upper limits on the bending strain effect. G could not be directly measured because values of α are small and difficult to measure at large bending diameters such as are being considered here. So an upper limit for α (and G) was obtained from measurements of $(I_c/I_{cm})_{uniaxial}$ obtained on the SRM material when it was subjected to <u>uniaxial</u> strain rather than bending strain.

Values of $(I_c/I_{cm})_{bending}$ were calculated from $(I_c/I_{cm})_{uniaxial}$ using the following equation applicable to a circular conductor [9]:

$$(I_c/I_{cm})_{bending} = \frac{2}{\pi \epsilon^2} \int_{-\epsilon_B}^{\epsilon_B} (\epsilon_B^2 - x^2)^{\frac{1}{2}} (I_c(x)/I_{cm})_{uniaxial} \, dx \tag{9}$$

In eq (9), $(I_c(x)/I_{cm})_{uniaxial}$ is the uniaxial strain degradation of I_c, assumed to be equal for both tensile and compressive strain, and ϵ_B is the peak bending strain in the outermost filaments. When a conductor is bent into an arc, the superconducting material on the outside of the bend is placed in tension while that on the inside of the bend is placed in compression. Assuming no yielding, the

13

material at the center of the conductor along the so-called "neutral axis" is not strained. The further a superconducting filament is from the neutral axis, the greater the bending strain it experiences. For a bend diameter of 3.23 cm (used to obtain the data reported here) and a SRM filament region diameter of 0.042 cm, the maximum bending strain, which occurs in the filaments nearest the surface of the conductor, is ε_B = 1.3%. G is obtained simply by evaluating eq (9) at this value of ε_B and then using eqs (7) and (8).

Values of $(I_c/I_{cm})_{uniaxial}$ used in eq (9) were measured on three specimens of the SRM material. In these measurements I_c/I_{cm} was determined to a precision of approximately 0.4%. The uniaxial strain degradation at 7 T is shown for three specimens in figure 4. It is seen that the effect of uniaxial strain on the SRM material falls within the shaded range of I_c degradation for a number of NbTi conductors having widely varying filament diameters and copper-to-superconductor ratios [10].

The uniaxial strain degradation increases with increasing magnetic field. Thus, the same relative increases with field would be expected for bending strain. Values of I_c/I_{cm} are shown in figure 5 at different magnetic fields for one of the specimens. As seen in figure 7, I_c/I_{cm} for a uniaxial strain of 1.3% is about 94.5% at 7 T, and increases to 96% at 1 T. The field-dependent upper limits on bending strain degradation listed at the end of the first paragraph of this section were calculated by substituting these measured values for $(I_c/I_{cm})_{uniaxial}$ into eq (9).

In eq (9) two major assumptions are made. First, it is assumed that differential yielding of the conductor material does not shift the neutral axis from the center of the conductor. Second, it is assumed that the conductor maintains a circular shape and there is no accommodative yielding of the

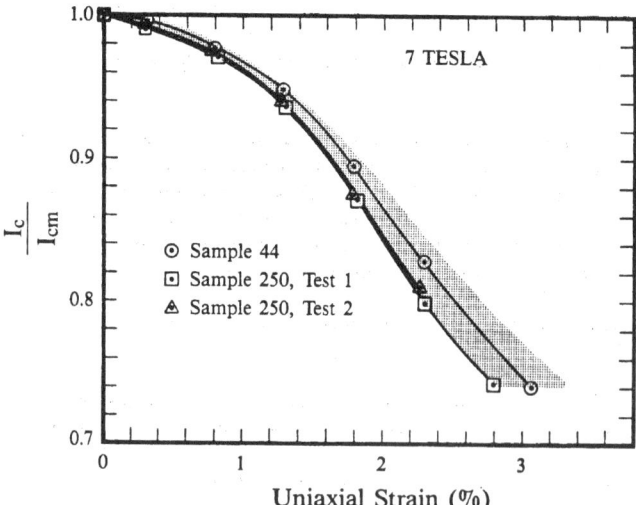

Figure 4. Uniaxial strain degradation of I_c for three SRM specimens at 7 T. The shaded region is the range of I_c degradation for a number of NbTi conductors.

14

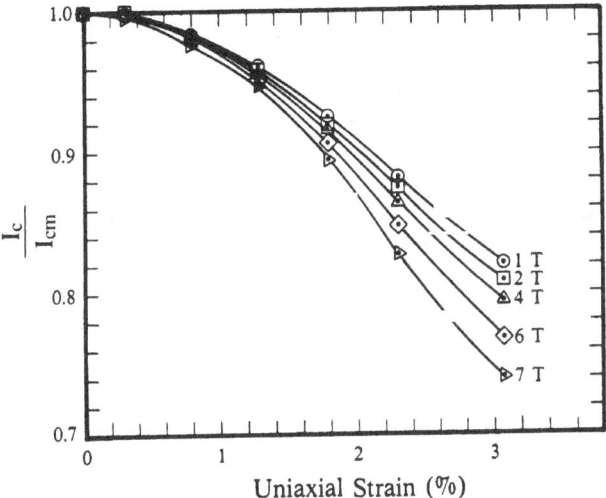

Figure 5. Uniaxial strain degradation of I_c for a SRM specimen at different magnetic fields.

soft copper matrix. In practice, both of these effects occur, but we believe that the second effect dominates (although no experimental evidence for this has been obtained yet). This would make the bending strain degradation calculated from eq (9) an upper limit for the actual room temperature bending degradation in I_c.

Bending strain effect measurements on two specimens support the theoretical upper limit to the value, eq (6), added to the overall uncertainty of the certified critical current. One of the specimens was measured straight (very large d) and the other with a bend diameter of 1.60 cm (approximately one-half the certified bend diameter). The upper limit for the uncertainty from bending determined from eq (6) is G for the straight specimen and 3.08 G for the bent specimen. In order to consider the conditions that could apply, the following variables are defined:

Δ = the absolute value of the difference between the measured critical current of the bent specimen and the certified value.

X = the absolute value of the worst-case limit to the estimated random error of the critical current measurement on the bent specimen (the sum of the random error terms in Table 1).

W = the absolute value of the estimated uncertainty limit of a single measurement due to material variability (see section 7.2).

15

Z = the absolute value of the added uncertainty due to the bending strain effect, using the theoretical expression eq (6), for specimens measured with bend diameters other than the bend diameter that the certification was based on.

Each variable is defined as an absolute value and in units of percent. The systematic error term does not enter for measurements made with the same apparatus. A very restrictive condition that can put an experimental upper limit on the effect of a certain bend is,

$$\Delta + X + W < Z. \tag{10}$$

Notice that this condition assumes the estimated upper limit to the material variability (worst-case variability) and thus need not be met by all measurements. If met by a single measurement, however, it supports the theoretical upper limit, providing the effect is well behaved (monotonic) over the range of bend diameters. This restrictive condition was satisfied by the 1.60-cm-diameter specimen at all four magnetic fields. This suggests that the theoretical limit may be conservative for bend diameters between 3.23 cm and 1.60 cm. The measurements on the straight specimen did not satisfy eq (10) because the inhomogeneity of the critical current was large compared to the expected bending effect; however, they were within the total uncertainty. These results give experimental support to the use of the theoretical upper limits to the bending strain effect and the allowed range of bend diameters.

3.9 Time

The time rate of change of the current, the ramp rate, may have an effect on the value of I_c. As stated in 2.3, the dependence was less than 0.1% for the range of ramp rates studied after the filtering and response time of the voltmeter are accounted for. The estimated systematic and random errors of this correction were included in the electric field variable. The suggested limits on the ramp rate are from zero to I_c in a time of from 300 s to 30 s, if the effect of the filtering and response time can be determined.

Mechanically and thermally cycling a specimen many times can change I_c. Mechanical cycling of a specimen by unmounting and remounting on a specimen holder is not recommended. The solder interfaces can concentrate the stress of handling, which will lower I_c in regions of the conductor. The possible effects of mechanical cycling were not tested because it would be difficult to test all of the contingencies, therefore the certification is not valid if the specimen is mechanically cycled. Similarly, thermal cycling is not recommended. It can introduce cumulative strain damage through static or dynamic (due to different cooling rates) differential thermal contraction if the elastic strain limit is exceeded. For this work, the magnet and helium dewar were pre-cooled with liquid nitrogen (76 K), then the liquid nitrogen was removed and the sample cryostat inserted. Thus, the cooling of the sample to liquid nitrogen temperature took place more slowly (with the gas heat exchange) than for an immersion into liquid nitrogen or liquid helium. With the sample cryostat in place, the liquid helium transfer slowly cooled the sample to 4 K. In a separate test, one specimen was measured, thermally cycled, and measured again. This specimen was cooled from room temperature to 4 K and the first measurements were taken. Subsequently, it was cycled three times from 4 K to approximately 70 K and then back to 4 K. It was also cycled once from 4 K to room temperature and back to 4 K where the specimen was measured again. The accumulative effect of this was less than a 0.2% change in I_c and it wasn't statistically significant compared to variations observed on repeat observations during a given run. The cumulative effect of thermal cycling is expected to be small, but this SRM cannot be certified beyond one thermal cycle. The certification cannot cover all thermal cycling possibilities.

4. PRELIMINARY SAMPLE SCREENING

The selection, from the five candidate conductors for the SRM, was based on the best balance of properties for use as an SRM. The chosen conductor may not be the best for any other application. A brief preliminary screening was designed to test two properties, the long-range (spool-to-spool) and the short-range (tap-to-tap) homogeneity of I_c. Two other key properties were availability of an adequate length and a usable copper-to-superconductor ratio. These four properties were sufficient to determine which conductor was chosen as the SRM.

4.1 Homogeneity and Other Key Properties

The homogeneity of I_c for each candidate was determined by measuring two specimens, located 5 and 50 meters from the end of the shipping spool. These two specimens are referred to in this section as the 5 m specimen and the 50 meter specimen respectively; their actual lengths were the same. Five pairs of voltage taps were placed on each specimen. Each pair of voltage taps spanned 2 cm and the centers of adjacent pairs were about 10 cm apart. Due to limitation of running time, all five pairs were measured only at 4 T. Several repeat measurements were made on one pair at the other magnetic fields. The results of this limited preliminary screening indicated homogeneity problems with two of the candidates and thus they were eliminated.

In no case were the above mentioned problems identifiable with a given or measured physical parameter. The particular parameters for each sample will not be identified. Only the values will be listed here, nonrespectively, for completeness. The wire diameters were: 0.40, 0.51, 0.51, 0.51, and 0.64 mm. The number of superconducting filaments were: 54, 54, 60, 126, and 180. The approximate superconducting filament diameters were: 23, 26, 34, 42, and 50 μm. The copper-to-superconductor ratios were: 1.4, 1.6, 1.8, 1.9, and 2.0 to 1.

The results of the tests of short and long range homogeneity for all five conductors are summarized in figures 6 and 7. Figure 6 is a plot of the percentage difference between I_c of the two specimens for each sample at various magnetic fields. Notice that the differences for each sample are about the same for the magnetic fields measured. Sample 4 had a large shift in I_c (approximately 3.9%). Another specimen from this spool, 150 m into the spool, was then measured. The I_c of this specimen had decreased another 1.1% from the value of the 50 m specimen (approximately 5.0% from the 5 m specimen). Lengths of wire adjacent to each of these specimens were tested to determine the mass of superconductor per unit length (nitric acid was used to remove the copper). The change from the 5 m specimens was 3.5% less for the 50 m specimen and 4.6% less for the 150 m specimen. The estimated uncertainty of this measurement was ±0.5%. The cross-sectional area changed by less than 0.5%. These measurements indicate that the change in I_c is due to a changing superconductor fraction. This may have been due only to an end effect, but this trend did not look good, so sample 4 was eliminated. Figure 7 is a plot of the percentage difference between the measured I_c on the center voltage pair of taps and that of the other four pairs of taps for both the 5 and 50 m specimen of each sample. I_c values determined for each tap pair has been corrected to account for the expected effect of the

17

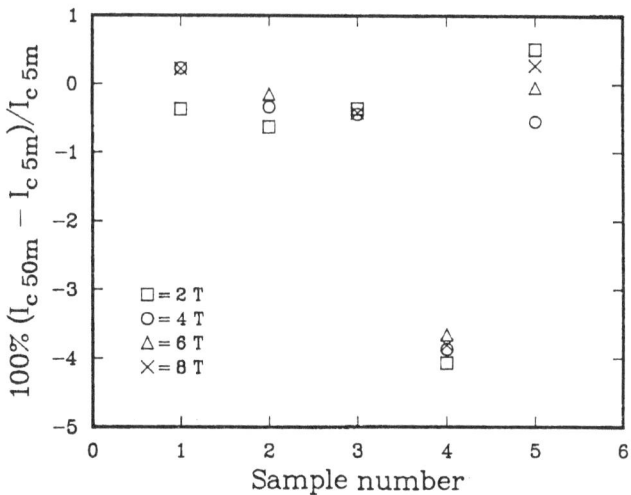

Figure 6. The percentage change in I_c between the 5 and 50 m
specimens of each sample, long-range homogeneity.

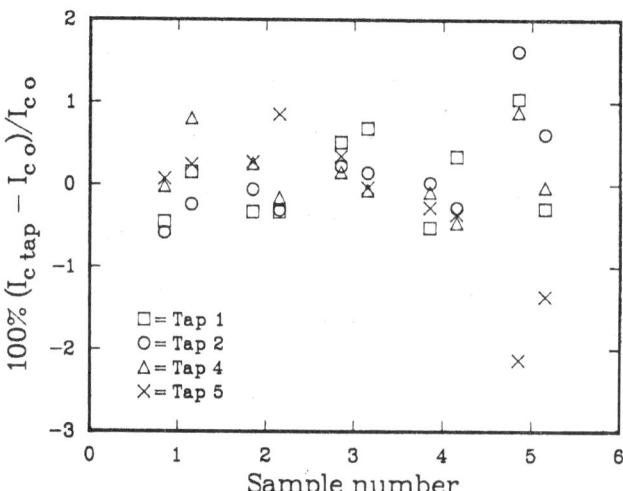

Figure 7. Short-range homogeneity of the 5 and 50 m specimens of each sample.
The 50 m data are shifted slightly to the right and the 5 m data to
the left over the sample number. Plotted are the percentage differences between the measured I_c on the center voltage tap and that of
the other four taps (at 4 T).

magnetic field profile. Notice that sample 5 seems to have substantially larger variations in this parameter and, for this reason, was eliminated.

The other two key properties were then used to select one conductor to be the SRM. One of the remaining candidates was eliminated because the length delivered was considered to be too short (less than 400 m) and an adjacent spool could not be obtained. One of the two candidates left had the lowest copper-to-superconductor ratio (1.4:1). It was eliminated even though the low ratio did not seem to adversely affect the I_c determination, but it could affect the usage of the SRM throughout the many kinds of testing anticipated for it.

4.2 SRM Spools

The SRM was manually unwound from the shipping spool, with a minimum bend diameter of about 12 cm onto distribution (take-up) spools with a core diameter of 8.7 cm. The shipping spool was mounted on a dowel wrapped with a fluorocarbon tape to minimize friction-caused tension. The distribution spool was mounted on a dowel in a fixture and a handle attached. The wire from the shipping spool was then pulled straight out and the end threaded into the distribution spool and affixed to a pin mounted on the handle. Eight turns were wound on the spool in a single layer corresponding to approximately 2.2 m of wire. Tension was applied by the operator's left hand while winding was done at a fairly uniform rate with the hand crank. As each spool was completed, a padded clip was used to retain the wire end, the wire was clipped and manually twisted to the end affixed to the pin. Each spool was then inspected, numbered, and any marks or stains noted. Upon further visual inspection, all of the marks or stains were only in the wire insulation, thus they should not have an effect on I_c. Special consideration was given to avoiding the introduction of any bending or twisting to the wire beyond what already existed on the shipping spool. One operator performed these operations on the entire set of samples, resulting in uniformity of handling of the set. Our qualifying tests were then performed on selected spools from this set.

Each SRM specimen was wound on a spool and packaged to protect its certification. The specimen should be carefully handled and stored to protect it against physical damage (such as excessive bending, scraping, and other deformation). Any excessive physical damage will invalidate the certification. On each spool, the twisted wire ends and an additional 2 cm on each end of the spool core should be discarded. These sections of the wire are not certified. The bending of the wire on the spool (peak bending strain of about 0.48%) is expected to be insignificant.

5. SAMPLE SPECIMENS AND MEASUREMENTS

Nine specimens were selected from the 500 spools of wire designated as SRM 1457. To assure that potential systematic trends in critical current could be detected, the nine sample spools were selected at nearly equal distances along the whole length of wire, including the spools at each end.

5.1 Range of Certified Values

SRM 1457 has been certified at each of four magnetic field strengths (2, 4, 6, and 8 T) over a range of temperatures from 3.90 to 4.24 K and a range of electric field criteria from 0.05 to 0.2 μV/cm.

19

The statistical analysis leading to the certification is based on measurements at 36 distinct combinations of the three factors affecting critical current:

Magnetic Field (H):   2, 4, 6, 8 T;
Temperature (T):   3.90, 4.07, 4.24 K;
Electric Field Criterion (E): 0.05, 0.10, 0.20 µV/cm.

Each of the nine sample spools were tested at 4.07 K for the 12 combinations of H and E. The 24 remaining configurations of the three factors, corresponding to 3.90 and 4.24 K, were tested on three of the sample specimens. The interpolation of certified values to cover the continuous range of T and E is based on a statistical model (discussed below) which has been constructed from the empirical relationship (see Section 2.2) for critical current, I_c, at a given magnetic field:

$$I_c = I_r [1 + \frac{T_r - T}{T_c^* - T_r}] \; [\frac{E}{E_r}]^{1/n} , \tag{11}$$

where T_r is a known reference temperature, E_r is a known reference electric field, and T_c^* is the unknown critical temperature at the selected magnetic field. Thus, if there are no errors, I_r is the critical current at (T_r, E_r) for a given H and eq (11) could be used to compute I_c for other combinations (T, E).

5.2 Measurement Procedure

The critical current data were obtained by placing three pairs of voltage taps on each specimen. The length of wire measured by each tap pair was about 2 cm and the centers of adjacent pairs were separated by about 20 cm. For a given magnetic field and temperature, digital voltage and current records acquired on each tap were analyzed to measure the critical current at various electric field criteria for both increasing and decreasing current. This procedure actually provides two values, I_c(up) and I_c(down), corresponding to increasing or decreasing current, respectively. However, the basic critical current measurements that were analyzed to certify SRM 1457 were defined to be the average:

$$I_c = \frac{I_c(up) + I_c(down)}{2}. \tag{12}$$

Two repeat determinations of I_c on each tap pair were obtained, producing a total of six measurements on a given spool (3 taps × 2 determinations) for each allowable level of (H, T, E).

5.3 Variation with Distance

Figures 8a-8d illustrate the critical current measurements, and variability, for the nine specimens of SRM 1457 when (T, E) = (4.07 K, 0.2 µV/cm) and H = 2, 4, 6, and 8 T. Both individual measurements and sample means are plotted in the figures. Figures 8a-8d adequately represent the general pattern of spool-to-spool variability which was observed at other levels (T, E). In particular, it

20

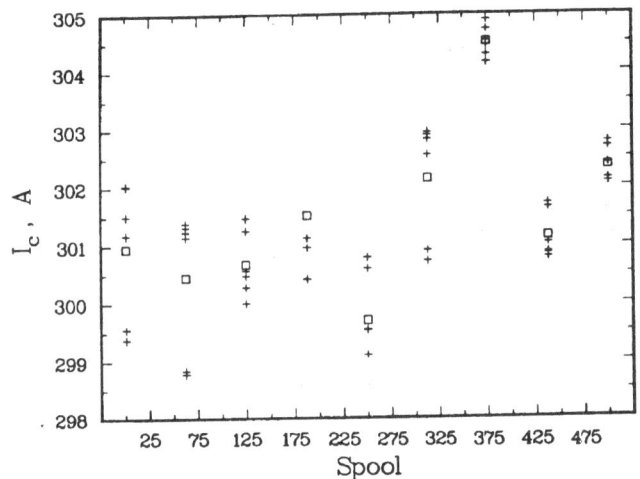

Figure 8a. Critical current measurements at 2 T, 4.07 K, and 0.2 μV/cm: I_c (+) and average I_c (□) versus distance along wire.

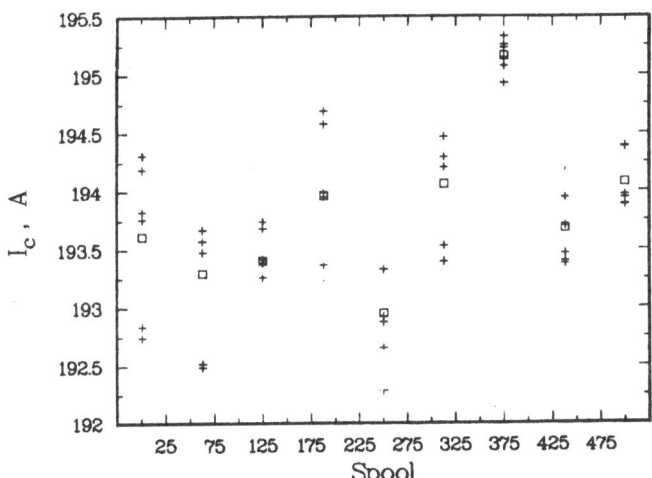

Figure 8b. Critical current measurements at 4 T, 4.07 K, and 0.2 μV/cm: I_c (+) and average I_c (□) versus distance along wire.

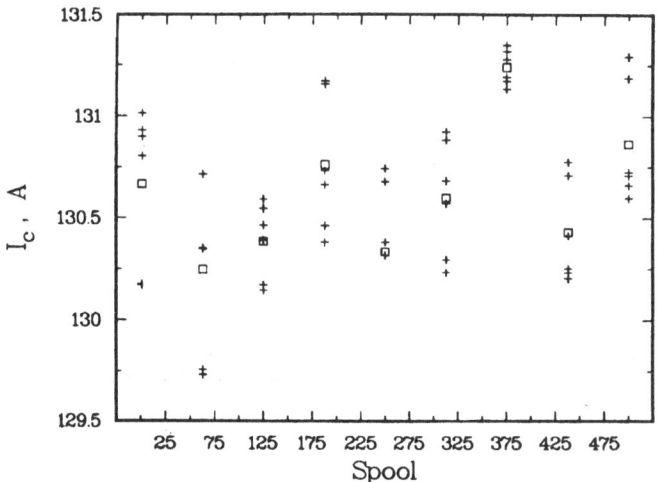

Figure 8c. Critical current measurements at 6 T, 4.07 K, and 0.2 μV/cm: I_c (+) and average I_c (□) versus distance along wire.

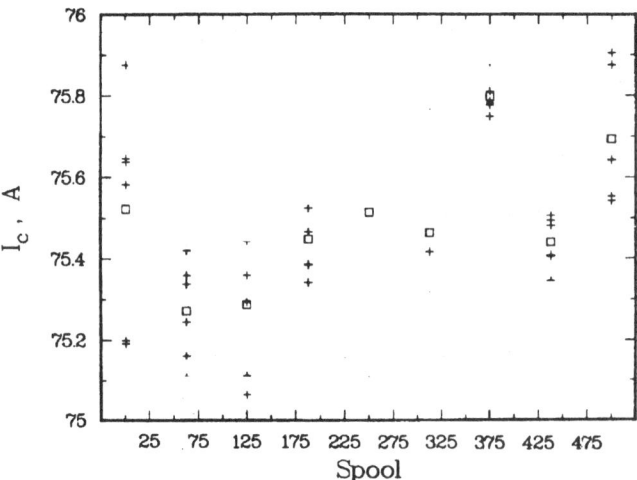

Figure 8d. Critical current measurements at 8 T, 4.07 K, and 0.2 μV/cm: I_c (+) and average I_c (□) versus distance along wire.

was concluded that there is no obvious trend in critical current along the length of wire -- distances being in one-to-one correspondence with assigned spool numbers.

Since no simple trend was detected, but the figures revealed substantial long- and short-range variations in critical current, it was decided that the 500 spools of SRM 1457 could not be assumed to have identical critical currents. Instead, the spool-to-spool and tap-to-tap variations in critical current illustrated by the figures have been incorporated into the statistical analysis for certification of SRM 1457 as randomly distributed offsets relative to the mean critical current of the 500 spools. The variations of these offsets have been associated with the long- and short-range material variability of the wire, and estimates of these variations are used in the certification to provide quantitative measures of inhomogeneity in the critical current SRM.

5.4 Variation with Magnetic Field

Typical variations of critical current with magnetic field are illustrated in figures 9a-9c. At each magnetic field, the measurements are expressed as a percentage of the average critical current. Each point is one of two determinations of I_c for each tap pair on each spool that was measured. Notice in the figures that variability of the measurements apparently depends on magnetic field, even if expressed as a percent of I_c.

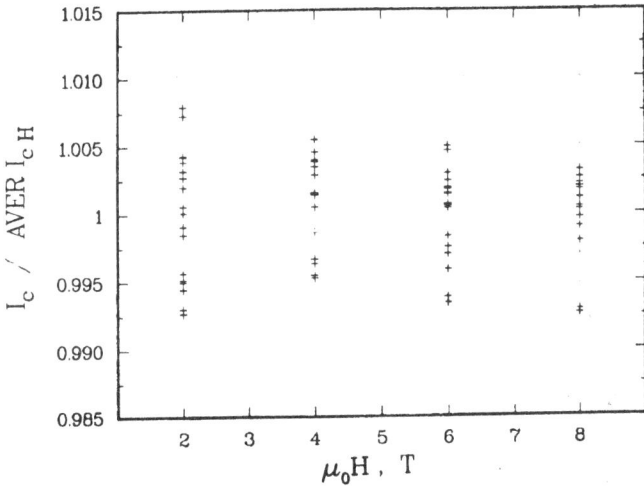

Figure 9a. Critical current measurements at 3.90 K and 0.2 μV/cm: (I_c/average I_c) versus magnetic field.

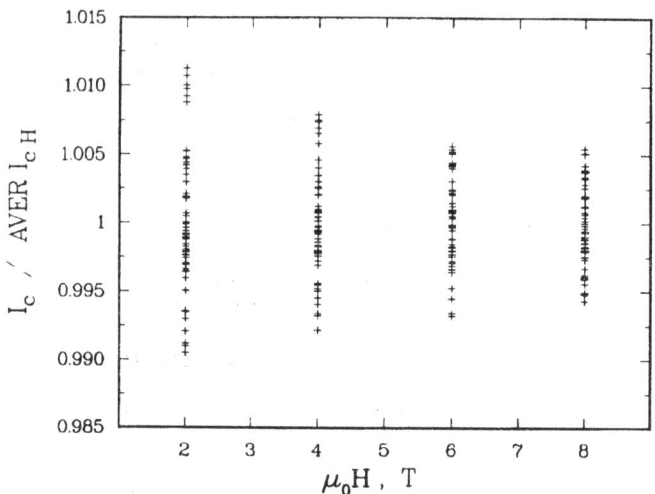

Figure 9b. Critical current measurements at 4.07 K and 0.2 µV/cm
(I_c/average I_c) versus magnetic field.

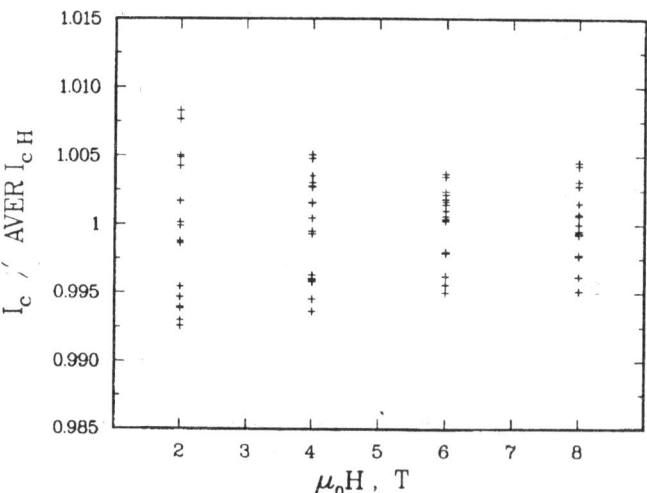

Figure 9c. Critical current measurements at 4.24 K and 0.2 µV/cm
(I_c/average I_c) versus magnetic field.

5.5 Variation with Temperature

Figures 10a–10d illustrate the variation of critical current with temperature at 2, 4, 6, and 8 T, respectively. The plots show that the observed relationship of critical current with temperature is consistent with the straight line predicted by the functional model in eq (11).

5.6 Variation with Electric Field

Figures 11a–11d illustrate the variation of critical current with electric field criteria at 2, 4, 6, and 8 T, respectively. Because of the form of the functional model in eq (11), both the sample data and electric field criteria were transformed to a logarithmic scale for the plots. Thus, the straight line dependence observed in the figures is consistent with the functional model.

5.7 Dependence on Electric Field and Temperature

The average critical current versus electric field criterion, with curves for each temperature, are shown in figures 12a–12d. The figures appear to confirm the curved, or exponential, relationship of critical current to electric field criteria at each fixed temperature. We also notice that relative differences between temperature averages are nearly fixed from one electric field criteria to another. In other words, the effects of temperature and electric field on critical current appear to be reasonably additive on the logarithmic scale.

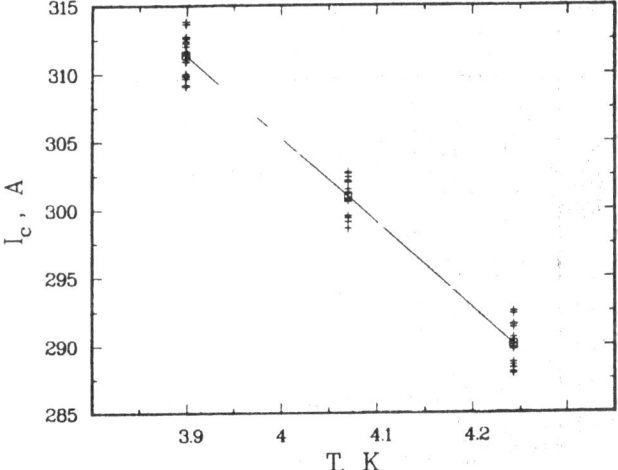

Figure 10a. Critical current measurements at 2 T and 0.2 μV/cm: I_c (+) and average I_c (□) versus temperature.

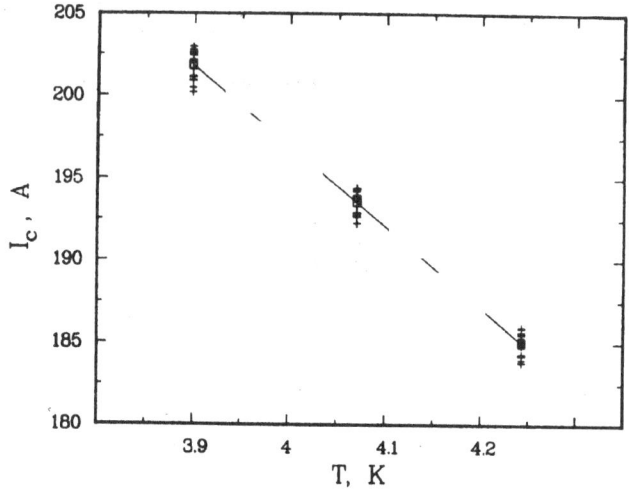

Figure 10b. Critical current measurements at 4 T and 0.2 μV/cm: I_c (+) and average I_c (□) versus temperature.

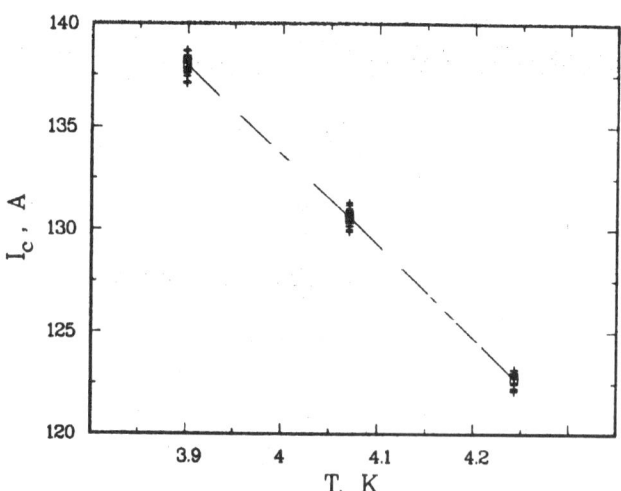

Figure 10c. Critical current measurements at 6 T and 0.2 μV/cm: I_c (+) and average I_c (□) versus temperature.

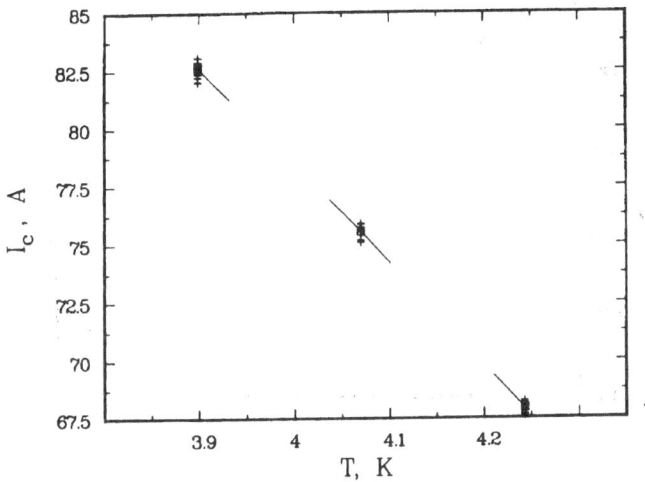

Figure 10d. Critical current measurements at 8 T and 0.2 μV/cm: I_c (+) and average I_c (□) versus temperature.

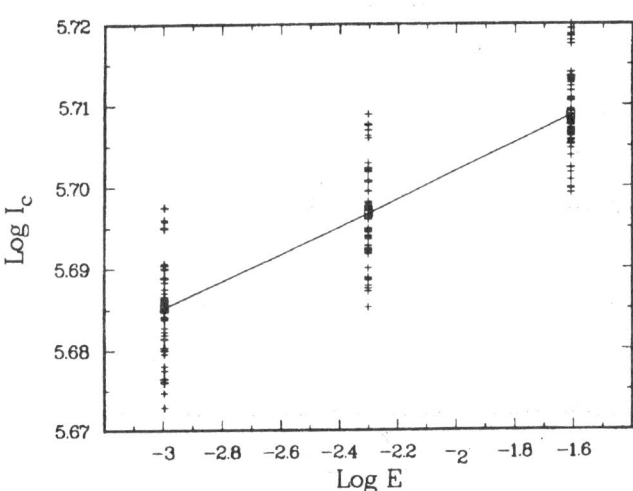

Figure 11a. Critical current measurements at 2 T and 4.07 K; natural log I_c (+) and average natural log I_c (□) versus natural log of electric field criterion.

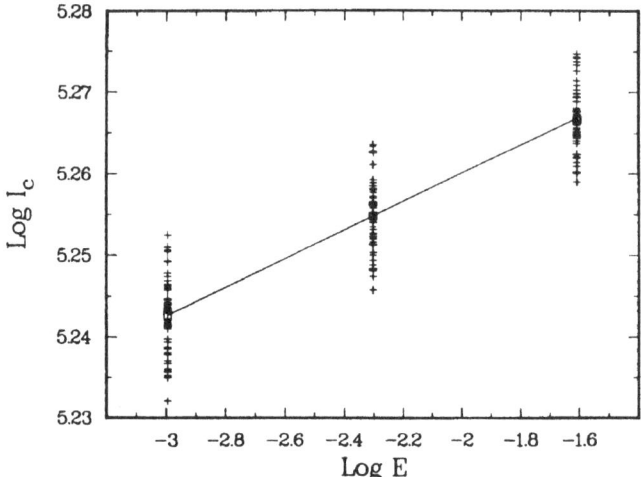

Figure 11b. Critical current measurements at 4 T and 4.07 K; natural log I_c (+) and average natural log I_c (□) versus natural log of electric field criterion.

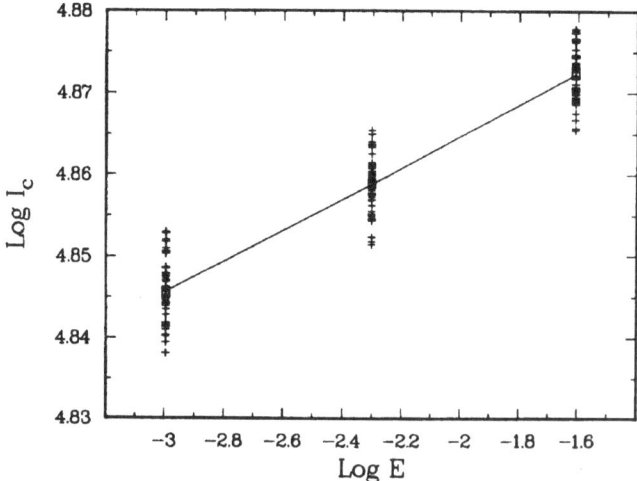

Figure 11c. Critical current measurements at 6 T and 4.07 K; natural log I_c (+) and average natural log I_c (□) versus natural log of electric field criterion.

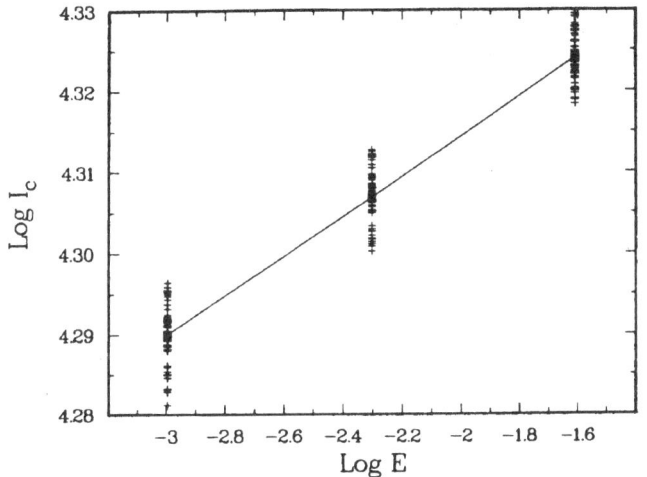

Figure 11d. Critical current measurements at 8 T and 4.07 K; natural log I_c (+) and average natural log I_c (□) versus natural log of electric field criterion.

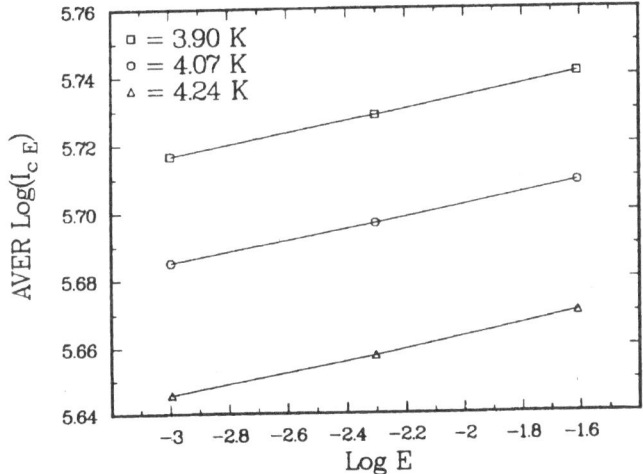

Figure 12a. Dependence of I_c on temperature and electric field at 2 T; points plotted are sample means of the natural logarithm of I_c.

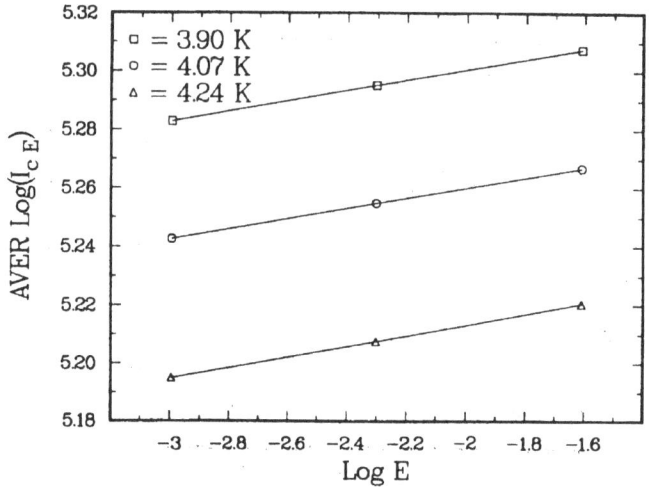

Figure 12b. Dependence of I_c on temperature and electric field at 4 T; points plotted are sample means of the natural logarithm of I_c.

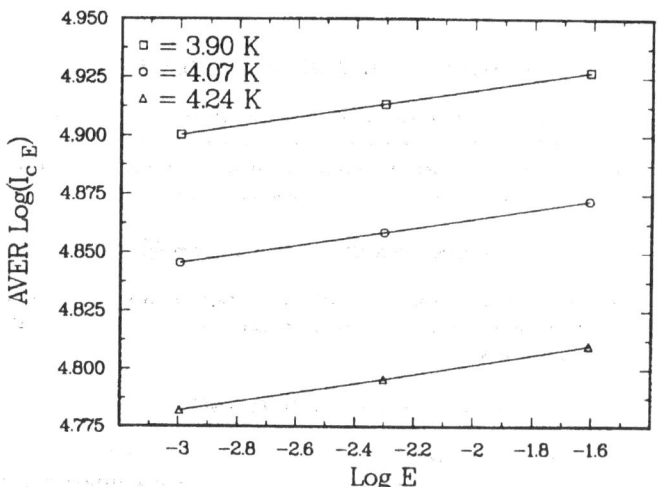

Figure 12c. Dependence of I_c on temperature and electric field at 6 T; points plotted are sample means of the natural logarithm of I_c.

30

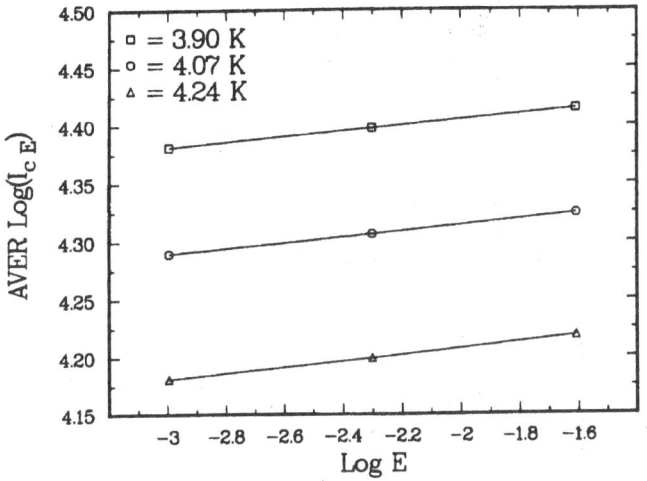

Figure 12d. Dependence of I_c on temperature and electric field at 8 T; points plotted are sample means of the natural logarithm of I_c.

5.8 Dependence on Magnetic Field and Temperature

Plots of average critical current versus magnetic field are shown in figures 13a–13c. Since the data at each magnetic field have been analyzed separately for certification, the behavior illustrated by the graphs has no bearing on subsequent statistical analysis. The graphs are included merely to describe the empirical dependence of critical current on magnetic field.

5.9 Dependence on Magnetic Field and Electric Field

Figures 14a–14c illustrates the dependence of critical current on magnetic field, with separate curves for each electric field criterion. The graphs are included for completeness and have no bearing on the statistical analysis.

6. ASSUMPTIONS AND STATISTICAL MODELS

The critical current data were first transformed to a logarithmic scale for three reasons. One reason is that the shape of the V-I curve suggests that critical current measurements may be nearly summable on a logarithmic scale. Secondly, empirical evidence that random measurement errors, and hence variation, are proportional to critical current suggests that the logarithmic data are more likely to satisfy underlying assumptions of the statistical methods. Finally, logarithmic transformation of temperature and electric field corrections in eq (11) allows us to express the statistical

31

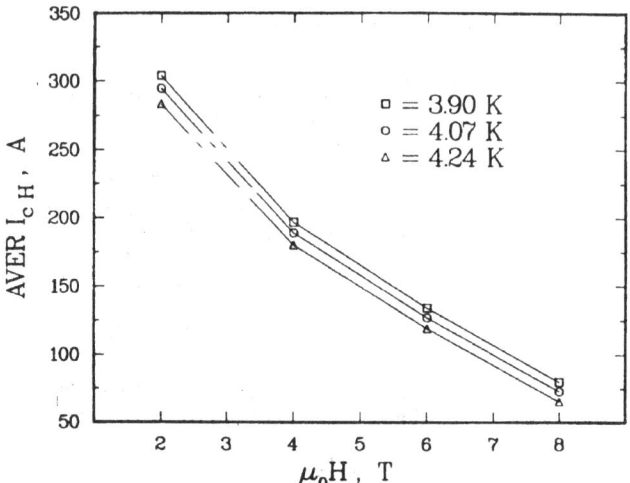

Figure 13a. Dependence of I_c on temperature and magnetic field at 0.05 μV/cm; points plotted are sample means.

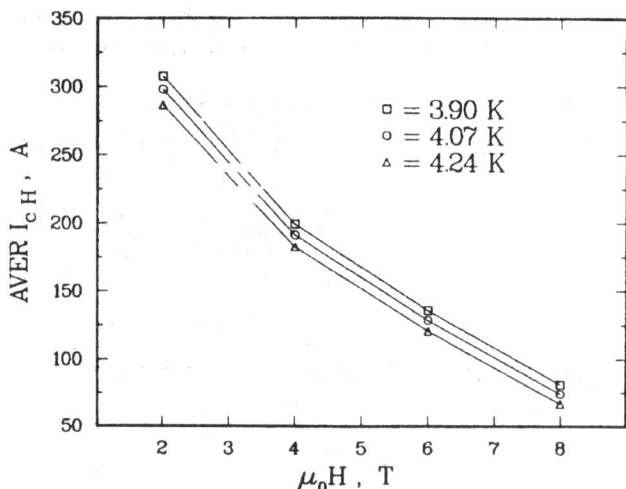

Figure 13b. Dependence of I_c on temperature and magnetic field at 0.10 μV/cm; points plotted are sample means.

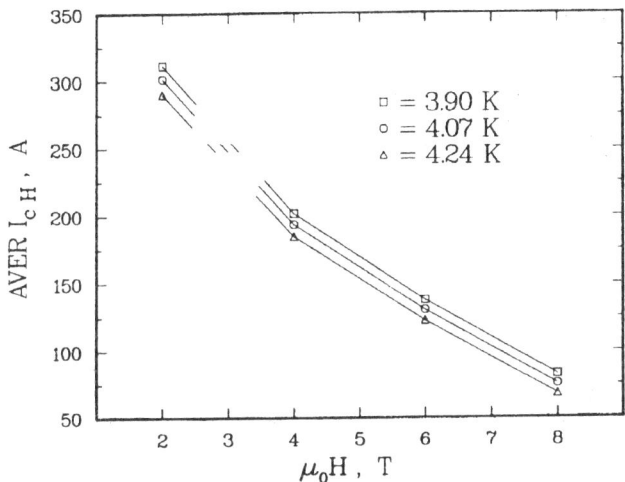

Figure 13c. Dependence of I_c on temperature and magnetic field at 0.20 µV/cm; points plotted are sample means.

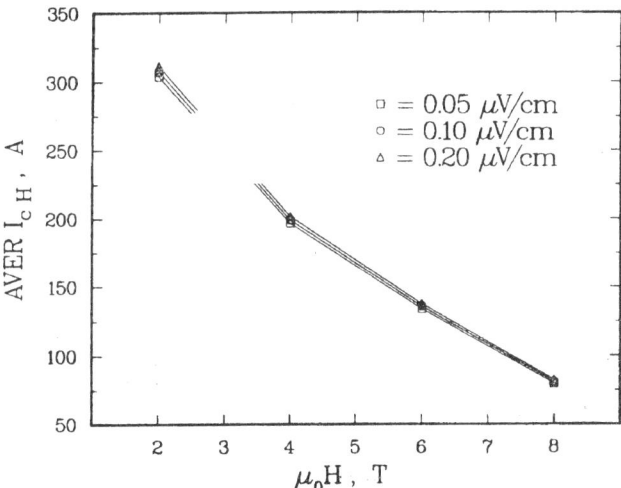

Figure 14a. Dependence of I_c on electric field and magnetic field at 3.90 K; points plotted are sample means.

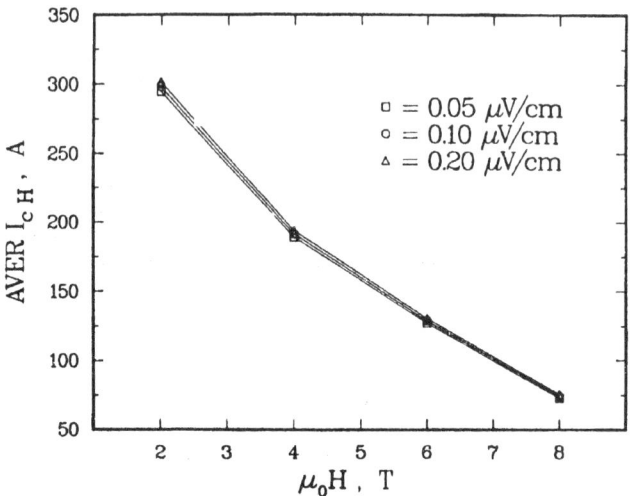

Figure 14b. Dependence of I_c on electric field and magnetic field at
4.07 K; points plotted are sample means.

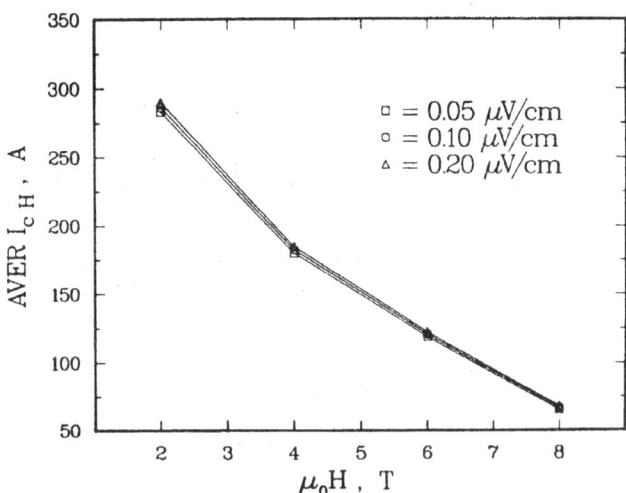

Figure 14c. Dependence of I_c on electric field and magnetic field at
4.24 K; points plotted are sample means.

model in a simple linear form. Therefore, all quantities of interest were estimated on the logarithmic scale, and then results on that scale have been converted to the original units of measurement.

6.1 Description of General Statistical Model

All of the critical current data at a given magnetic field were incorporated into a common statistical model derived from the approximate relationship of I_c to temperature (T) and electric field (E), and based on the observation that individual spools and/or taps exhibit unpredictable, random differences from the average critical current of all spools. Thus, each measurement can be considered to be a composite of the empirical relationship in eq (11) and the random effects of material inhomogeneity. In terms of the natural logarithm of critical current, each measurement at a given magnetic field can be thought of as the sum of six components:

$$Y_{ijk}(T_\ell, E_m) = \mu_r + \ell n[1 + \frac{T_r - T_\ell}{T_c^* - T_r}] + n^{-1} \ell n[\frac{E_m}{E_r}] + D_i + L_{j(i)} + e_{ijk\ell m}. \tag{13}$$

If, for convenience, we let τ_ℓ and ε_m represent the functions in the model involving temperature and electric field criterion, then an explanation of terms in the model is given by:

$Y_{ijk}(T_\ell, E_m)$ is the natural logarithm of I_c for the kth repeat determination on the jth tap on the ith spool when the temperature is set at T_ℓ and the electric field criteria is E_m;

μ_r is the natural logarithm of the mean critical current at the reference point (T_r, E_r);

τ_ℓ is a correction to μ_r if $T_\ell \neq T_r$;

ε_m is a correction to μ_r if $E_m \neq E_r$;

D_i is a random (long-range) inhomogeneity of the ith spool, corresponding to distance along the wire; the D's are assumed to have average value zero and standard deviation σ_D;

$L_{j(i)}$ is the random (short-range) inhomogeneity, corresponding to the location of the jth tap on the ith spool; the L's are assumed to have average value zero and standard deviation σ_L;

$e_{ijk\ell m}$ is the random error in the measurement $Y_{ijk}(T_\ell, E_m)$; the e's are assumed to have average value zero and standard deviation σ.

6.2 Approximation of Temperature Dependence

To facilitate the estimation of the unknown quantities, the temperature correction term, which is a nonlinear function of T_c^*, has been approximated using the Taylor series relation:

$$\ell n\, x = (x - 1) - \tfrac{1}{2}(x - 1)^2 + \text{Remainder}, \quad 0 < x \leq 2. \tag{14}$$

35

If the reference temperature is selected near the center of the range 3.90 to 4.24 K, then it is reasonable to disregard terms in the expansion higher than second order for values of T_c^* which can be expected at magnetic fields from 2 to 8 T.

6.3 Statistical Model for Sample Data

Incorporating the approximation for the temperature correction into the model and substituting the reference values $(T_r, E_r) = (4.07 \text{ K}, 0.2 \text{ } \mu\text{V/cm})$, which were selected for the statistical analysis, we can write:

$$Y_{ijk}(T_\ell, E_m) = \mu + \beta_1(4.07 - T_\ell) + \beta_2(4.07 - T_\ell)^2 + \beta_3 \ln(E_m/0.2) + D_i + L_{j(i)} + e_{ijk\ell m} \quad (15)$$

where,

$$\beta_1 = (T_c^* - 4.07)^{-1}$$

$$\beta_2 = -0.5(T_c^* - 4.07)^{-2} \quad (16)$$

and $$\beta_3 = n^{-1}.$$

For each set of data at 2, 4, 6, or 8 T, estimates of μ, β_1, β_2, β_3, σ_D, σ_L, and σ were computed using the statistical software package BMDP [11]. The data were fit by a maximum likelihood procedure described in the program BMDP3V. The statistical analysis also provides estimated standard errors of the fitted values, which were used to determine uncertainty bounds on both the certified critical current and the estimates of T_c^* and n derived from eq (16).

7. STATISTICAL ANALYSIS AND CERTIFICATION

7.1 Material Inhomogeneity

An important question for certification is whether all the available spools for this SRM can be treated as identical. The components of variance associated with the long- and short-range material variability of the wire provide a quantitative measure of inhomogeneity in the critical current SRM, so estimates of each type of variation were obtained. The estimated standard deviations, expressed as a percentage error in the critical current at each of the magnetic fields, are given in table 2. The table illustrates that long-range inhomogeneity is more evident at 2 T and decreases with increasing magnetic field. Estimates of standard deviations of repeat determinations of critical current, and the total standard deviation of a single measurement on a randomly selected spool are also given in table 2.

Because the critical current measurements revealed substantial long- and short-range variation, the uncertainty statement for the reported critical current at each magnetic field is a statistical tolerance interval. This statistical procedure allows for the observed variation in critical current by estimating limits for the critical current of individual spools, rather than limits on the overall average critical current of all spools.

Table 2. Estimated Long- and Short-Range Inhomogeneity Expressed in Percent Error in I_c.

Magnetic Field (T)	Spool-to-Spool Std. Dev. (σ_D)	Taps within Spools Std. Dev. (σ_L)	Total Inhomogeneity $(\sigma_D^2 + \sigma_L^2)^{\frac{1}{2}}$	Repeat Determination Std. Dev. (σ)	Total Std. Dev. $(\sigma_D^2 + \sigma_L^2 + \sigma^2)^{\frac{1}{2}}$
2	0.46%	0.35%	0.57%	0.08%	0.58%
4	0.33	0.29	0.44	0.09	0.45
6	0.22	0.27	0.35	0.09	0.36
8	0.16	0.27	0.31	0.15	0.35

7.2 Tolerance Limits for Critical Current

Tolerance limits for the distribution of critical currents for a length of wire of about 2 cm have been estimated from the sample data by the formula:

$$\exp[\hat{Y}(T, E) \pm K(\hat{\sigma}_D^2 + \hat{\sigma}_L^2 + \hat{\sigma}^2)^{\frac{1}{2}}], \tag{17}$$

where

$$\hat{Y}(T, E) = \hat{\mu} + \hat{\beta}_1(4.07 - T) + \hat{\beta}_2(4.07 - T)^2 + \hat{\beta}_3 \ln(E/0.2).$$

A circumflex (^) denotes computed estimates of the various quantities. The constant K is usually taken from tables, such as in Weissberg and Beatty [12]. The value of K depends on four quantities:

(1) N: the effective number of observations for $\hat{Y}$;
(2) f: the degrees of freedom associated with $(\hat{\sigma}_D^2 + \hat{\sigma}_L^2 + \hat{\sigma}^2)$;
(3) P: the proportion of critical current values to be included in the limits; and,
(4) v: the probability level associated with the tolerance interval.

The tolerance limits for SRM 1457 are approximate because exact values of N and f are not known for the model in eq (15). Instead, N was estimated from the data as the integer part of $[(\hat{\sigma}_D^2 + \hat{\sigma}_L^2 + \hat{\sigma}^2)/$ Estimated Variance of $\hat{Y}(T, E)]$, and f was estimated using Satterthwaite's procedure [13]. It is important to note that, in general, the tolerance limits depend on the temperature and electric field criteria because N, and hence K, depend on the estimated variance of $\hat{Y}(T, E)$. However, for the observed levels of inhomogeneity of SRM 1457, the tolerance limits, when expressed in percent error in I_c, do not change for temperatures from 3.90 to 4.24 K and electric field criteria from 0.05 to 0.2 µV/cm.

7.3 Certified Critical Current at 4.2 K and 0.2 µV/cm

The estimated superconducting critical currents at 4.2 K for an electric field criterion 0.2 µV/cm for a tap spacing of 2 cm are given in table 3. The uncertainty of the reported value, ignoring systematic errors, is the statistical tolerance interval constructed such that it should cover 99% of critical current determinations with probability 0.95. The tolerance limits in table 3 are expressed

Table 3. Critical Current at 4.2 K and 0.2 µV/cm.

Magnetic Field (T)	Critical Current (A)	Total Uncertainty (%)	Tolerance Limits (%)	Systematic Error (%)
2.000	293.30†	2.57	2.33	0.24
4.000	187.38	2.01	1.74	0.27
6.000	124.72	1.71	1.32	0.39
8.000	69.72	1.97	1.19	0.78

†Extra digits are provided for accurate extrapolation.

in terms of the resultant percentage error at each magnetic field and apply to a single measurement on any given spool for a length of wire of about 2 cm. Uncertainties for wire lengths greater than 2 cm are expected to be less than those in table 3. The estimated systematic error and the tolerance limit are summed to give the total uncertainty and are all expressed as percent error in I_c.

7.4 Temperature and Electric Field Corrections

Certified critical currents for temperatures (T) between 3.90 and 4.24 K and electric field criteria (E) between 0.05 and 0.2 µV/cm are computed from values in table 3 using the equation:

$$I_c(T,E) = I_c(4.2,0.2) \cdot (\exp[A(4.2-T) + B(4.2-T)^2]) \cdot [E/0.2]^C , \qquad (18)$$

where $I_c(4.2,0.2)$ is the tabulated value and $A = \hat{\beta}_1 - 0.26\hat{\beta}_2$, $B = \hat{\beta}_2$, and $C = \hat{\beta}_3$. Uncertainty in a critical current computed this way is the same as in table 3. The peculiar form of A arises because temperatures were centered about 4.07 K to estimate coefficients in eq (15), but it was decided to report critical currents at 4.2 K on the certificate for the SRM.

The coefficients required in eq (18) are listed in table 4. Certified critical currents computed by substituting the coefficients into eq (18) are given in table 5 for three electric field criteria at temperatures from 3.90 to 4.24 K in 0.01 K increments.

Table 4. Coefficients for Temperatures and Electric Field Corrections.

Magnetic Field (T)	A	B	C
2.000	0.218625	−0.04755	0.0172089
4.000	0.266361	−0.04682	0.0176600
6.000	0.369479	−0.10488	0.0194218
8.000	0.649242	−0.27906	0.0248311

Table 5. Critical Current at 0.05, 0.10, 0.20 µV/cm and 3.90, 3.91, ..., 4.24 K.

3.90	2	304.49	308.15	311.84	2.57
	4	197.23	199.66	202.11	2.01
	6	134.36	136.18	138.03	1.71
	8	79.82	81.20	82.61	1.97
3.91	2	303.91	307.56	311.25	2.57
	4	196.76	199.18	201.63	2.01
	6	133.95	135.77	137.61	1.71
	8	79.43	80.81	82.21	1.97
3.92	2	303.33	306.97	310.66	2.57
	4	196.28	198.70	201.15	2.01
	6	133.54	135.35	137.18	1.71
	8	79.04	80.41	81.81	1.97
3.93	2	302.75	306.38	310.06	2.57
	4	195.81	198.22	200.67	2.01
	6	133.12	134.93	136.75	1.71
	8	78.65	80.02	81.41	1.97
3.94	2	302.16	305.79	309.46	2.57
	4	195.34	197.75	200.18	2.01
	6	132.70	134.50	136.33	1.71
	8	78.26	79.62	81.00	1.97
3.95	2	301.58	305.20	308.86	2.57
	4	194.87	197.27	199.70	2.01
	6	132.29	134.08	135.90	1.71
	8	77.86	79.21	80.59	1.97
3.96	2	300.99	304.60	308.25	2.57
	4	194.39	196.79	199.21	2.01
	6	131.87	133.65	135.46	1.71
	8	77.46	78.81	80.18	1.97
3.97	2	300.40	304.00	307.65	2.57
	4	193.92	196.31	198.73	2.01
	6	131.44	133.23	135.03	1.71
	8	77.06	78.40	79.76	1.97
3.98	2	299.81	303.40	307.04	2.57
	4	193.44	195.83	198.24	2.01
	6	131.02	132.80	134.60	1.71
	8	76.66	77.99	79.35	1.97
3.99	2	299.21	302.80	306.44	2.57
	4	192.97	195.35	197.75	2.01
	6	130.60	132.37	134.16	1.71
	8	76.26	77.58	78.93	1.97
4.00	2	298.62	302.20	305.83	2.57
	4	192.49	194.86	197.26	2.01
	6	130.17	131.94	133.72	1.71
	8	75.85	77.17	78.51	1.97
4.01	2	298.02	301.60	305.22	2.57
	4	192.02	194.38	196.77	2.01
	6	129.74	131.50	133.28	1.71
	8	75.44	76.75	78.08	1.97
4.02	2	297.42	300.99	304.60	2.57
	4	191.54	193.90	196.28	2.01
	6	129.32	131.07	132.84	1.71
	8	75.03	76.33	77.66	1.97

Table 5 (Continued)

				:m)	Total
				).20	Uncertaint
4.03	2	296.82	300.38	303.99	2.57
	4	191.06	193.41	195.79	2.01
	6	128.89	130.63	132.40	1.71
	8	74.62	75.91	77.23	1.97
4.04	2	296.22	299.77	303.37	2.57
	4	190.58	192.93	195.30	2.01
	6	128.45	130.20	131.96	1.71
	8	74.20	75.49	76.80	1.97
4.05	2	295.62	299.16	302.75	2.57
	4	190.10	192.44	194.81	2.01
	6	128.02	129.76	131.52	1.71
	8	73.79	75.07	76.37	1.97
4.06	2	295.01	298.55	302.13	2.57
	4	189.62	191.96	194.32	2.01
	6	127.59	129.32	131.07	1.71
	8	73.37	74.64	75.94	1.97
4.07	2	294.41	297.94	301.51	2.57
	4	189.14	191.47	193.83	2.01
	6	127.15	128.88	130.63	1.71
	8	72.95	74.21	75.50	1.97
4.08	2	293.80	297.32	300.89	2.57
	4	188.66	190.98	193.34	2.01
	6	126.72	128.44	130.18	1.71
	8	72.53	73.79	75.07	1.97
4.09	2	293.19	296.71	300.27	2.57
	4	188.18	190.50	192.84	2.01
	6	126.28	127.99	129.73	1.71
	8	72.10	73.36	74.63	1.97
4.10	2	292.58	296.09	299.64	2.57
	4	187.70	190.01	192.35	2.01
	6	125.84	127.55	129.28	1.71
	8	71.68	72.92	74.19	1.97
4.11	2	291.96	295.47	299.01	2.57
	4	187.21	189.52	191.85	2.01
	6	125.41	127.10	128.83	1.71
	8	71.25	72.49	73.75	1.97
4.12	2	291.35	294.85	298.38	2.57
	4	186.73	189.03	191.36	2.01
	6	124.97	126.66	128.38	1.71
	8	70.83	72.05	73.31	1.97
4.13	2	290.73	294.22	297.75	2.57
	4	186.25	188.54	190.86	2.01
	6	124.52	126.21	127.92	1.71
	8	70.40	71.62	72.86	1.97
4.14	2	290.12	293.60	297.12	2.57
	4	185.76	188.05	190.37	2.01
	6	124.08	125.76	127.47	1.71
	8	69.97	71.18	72.42	1.97
4.15	2	289.50	292.97	296.49	2.57
	4	185.28	187.56	189.87	2.01
	6	123.64	125.31	127.01	1.71
	8	69.53	70.74	71.97	1.97

Table 5 (Continued)

Temperature (K)	Magnetic Field (T)	Electric Field Criterion (µV/cm)			Total Uncertainty (%)
		0.05	0.10	0.20	
4.16	2	288.88	292.35	295.85	2.57
	4	184.79	187.07	189.37	2.01
	6	123.19	124.86	126.56	1.71
	8	69.10	70.30	71.52	1.97
4.17	2	288.26	291.72	295.22	2.57
	4	184.31	186.58	188.88	2.01
	6	122.75	124.41	126.10	1.71
	8	68.67	69.86	71.07	1.97
4.18	2	287.64	291.09	294.58	2.57
	4	183.82	186.09	188.38	2.01
	6	122.30	123.96	125.64	1.71
	8	68.23	69.42	70.62	1.97
4.19	2	287.01	290.46	293.94	2.57
	4	183.34	185.59	187.88	2.01
	6	121.85	123.51	125.18	1.71
	8	67.80	68.97	70.17	1.97
4.20	2	286.39	289.82	293.30	2.57
	4	182.85	185.10	187.38	2.01
	6	121.41	123.05	124.72	1.71
	8	67.36	68.53	69.72	1.97
4.21	2	285.76	289.19	292.66	2.57
	4	182.36	184.61	186.88	2.01
	6	120.96	122.60	124.26	1.71
	8	66.92	68.08	69.27	1.97
4.22	2	285.13	288.55	292.01	2.57
	4	181.87	184.11	186.38	2.01
	6	120.51	122.14	123.80	1.71
	8	66.48	67.64	68.81	1.97
4.23	2	284.50	287.92	291.37	2.57
	4	181.39	183.62	185.88	2.01
	6	120.06	121.68	123.33	1.71
	8	66.04	67.19	68.36	1.97
4.24	2	283.87	287.28	290.72	2.57
	4	180.90	183.12	185.38	2.01
	6	119.61	121.23	122.87	1.71
	8	65.60	66.74	67.90	1.97

7.5 Critical Temperature and Exponents

If $\hat{\beta}_1$, $\hat{\beta}_2$, and $\hat{\beta}_3$ denote estimates of the parameters, and $S(\hat{\beta}_1)$, $S(\hat{\beta}_2)$ and $S(\hat{\beta}_3)$ are computed standard errors of the estimates, then n and T_c^* can be estimated by the equations:

$$\hat{n} = \hat{\beta}_3^{-1}, \quad S(\hat{n}) = S(\hat{\beta}_3)/\hat{\beta}_3^2$$

and

$$\hat{T}_c^* = T_r + \frac{w_1 \hat{\beta}_1^{-1} + w_2(-2\hat{\beta}_2^{-2})}{w_1 + w_2}, \quad s(\hat{T}_c^*) = (w_1 + w_2)^{-\frac{1}{2}} \tag{19}$$

where

$$w_1 = [s^2(\hat{\beta}_1)/\hat{\beta}_1^4]^{-1} \quad \text{and} \quad w_2 = [16 \; s^2(\hat{\beta}_2)/\hat{\beta}_2^6]^{-1}.$$

The solution for T_c^* is a weighted estimate, where the weights, w_1 and w_2, are inverses of the estimated variances for separate estimates of T_c^* obtained from $\hat{\beta}_1$ and $\hat{\beta}_2$, respectively. While all of
the formulas for standard deviations are approximate, each was derived using standard propagation of
error techniques [14]. Also, the expression for $S(\hat{T}_c^*)$ is appropriate only if estimates of β_1 and β_2
are statistically independent. The two estimates are approximately independent in the present analysis.

Estimates of T_c^* and n and their computed standard deviations are listed in table 6. Using the
values of T_c^* and n in table 5 and the critical current at 4.2 K and 0.2 µV/cm reported in table 3,
the values of critical current for a temperature (T) between 3.90 and 4.24 K and electric field
criteria (E) between 0.05 and 0.2 µV/cm could be computed from:

$$I_c(T, E) = I_c(4.2, 0.2)[1 + \frac{4.2 - T}{T_c^* - 4.2}] \; [\frac{E}{0.2}]^{1/n} . \tag{20}$$

However, critical currents computed this way will differ slightly from the certified values because
the temperature correction term above is not equivalent (even to second order terms) with the corresponding term in eq (18). The numerical (or systematic) discrepancy between the two equations can be
as large as 0.2% at 8 T.

7.6 Additional Data Excluded from Certification

Additional measurements of critical current were made on the nine sample specimens at electric
field criteria 0.02 and 0.5 µV/cm. However, SRM 1457 has not been certified to the extended range of
electric field criteria because measurements at 0.02 µV/cm were made only for increasing current, and
some of the data at 0.5 µV/cm were censored when the ramp was reversed before this criterion was
reached.

Table 6. Critical Temperature (T_c^*) and Exponent (n).

Magnetic Field (T)	T_c^* (K)	Estimated Standard Deviation (K)	n	Estimated Standard Deviation
2.000	8.9182	0.0105	58.11	0.29
4.000	8.0041	0.0080	56.63	0.32
6.000	6.9922	0.0041	51.49	0.25
8.000	5.8040	0.0026	40.27	0.27

Table 7. Critical Currents at 0.02 µV/cm.

Tempera-ture (K)	Magnetic Field (T)	Number Observed	Sample Mean (A)	Predicted Value (A)	Tolerance Limits (%)	Number Within Limits (%)
3.89900	2	18	299.72	299.78	2.33	100
	4	18	194.20	194.11	1.74	100
	6	18	132.28	132.04	1.32	100
	8	18	78.47	78.06	1.19	94
4.06995	2	54	290.38	289.80	2.33	100
	4	54	186.58	186.11	1.74	100
	6	54	125.25	124.92	1.32	100
	8	54	71.51	71.31	1.19	100
4.24273	2	18	279.14	279.26	2.33	100
	4	18	177.79	177.87	1.74	100
	6	18	117.50	117.38	1.32	100
	8	18	64.01	64.01	1.19	100

The available critical current data at 0.02 and 0.5 µV/cm have been compared to values extrapolated from eq (18) for information only. Tables 7 and 8 list the observed average critical current, the critical current extrapolated from the certified values, and the percentage of measurements at 0.02 and 0.5 µV/cm that are within the range computed using tolerance limits which are valid between 0.05 and 0.2 µV/cm. Data at 0.5 µV/cm were included in table 8 only if critical currents were observed for both increasing and decreasing current.

Figures 15a-15d illustrate the dependence of critical current on electric field criterion from 0.02 to 0.5 µV/cm. The values plotted are sample means of the logarithms of critical current measurements on the sample specimens.

8. DISCUSSION

The observed variation in the critical current was probably due to inhomogeneity in the NbTi alloy. This is suggested by the fact that the tolerance limit at 8 T is about half the value at 2 T. This indicates that a number of sources of variation that would have no field dependence or a field dependence in the opposite sense (i.e., a larger variation at 8 T than at 2 T) must not be present or are only present at a minor level. The sources ruled out are variation in the following: specimen handling, the upper critical field, the critical temperature, the amount of superconductor, and filament size. The most likely source of the larger variation at 2 T is a variation in pinning strength. A variation in pinning strength can be caused by variation in the following: heat treatment, cold work, impurity density, and alloy concentration. Since the wire was supplied as a single length (2.2 km) and the variations were observed to be short-range (20 cm) as well, a variation in heat treatment or cold work is probably not the source. Inhomogeneity of the alloy impurities and Nb concentration are the most likely sources of the observed variation in the critical current.

An effort was made to keep the use of this SRM as unrestricted as possible. The precautions listed on the certificate together with the ASTM standard test method are sufficient for a valid user measurement technique. Some deviations in testing technique from the certification based method are accommodated by increasing the total uncertainty of the certified critical current. The deviations

Table 8. Critical Currents at 0.5 μV/cm.

Tempera-ture (K)	Magnetic Field (T)	Number Observed	Sample Mean (A)	Predicted Value (A)	Tolerance Limits (%)	Number Within Limits (%)
3.89900	2	13	316.28	316.86	2.33	100
	4	9	204.91	205.46	1.74	100
	6	12	140.38	140.55	1.32	100
	8	3	84.07	84.55	1.19	100
4.06995	2	33	306.12	306.30	2.33	100
	4	16	196.91	197.00	1.74	100
	6	20	132.96	132.97	1.32	100
	8	13	77.16	77.24	1.19	100
4.24273	2	18	295.19	295.16	2.33	100
	4	8	187.88	188.27	1.74	100
	6	10	125.05	124.95	1.32	100
	8	6	69.80	69.34	1.19	100

that are allowed, and the ones that are not allowed, are identified in the precautions section of the certificate. Two deviations that are allowed at some sacrifice to the uncertainty are rapid (immersive) cooling of the specimen and measurements with different bend diameters. Another precaution is that the certification is only valid for a voltage tap separation greater than or equal to 2 cm. This restriction was a result of the short-range inhomogeneity of the SRM. The critical current of the specimen is not certified beyond the first mounting, cool down, and measurement (the first mechanical and thermal cycle).

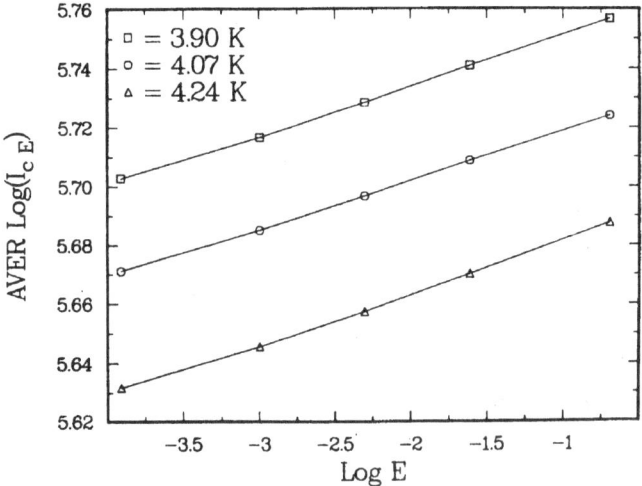

Figure 15a. Dependence of I_c on electric fields from 0.02 to 0.5 μV/cm at 2 T; natural logarithm of I_c and electric field are plotted.

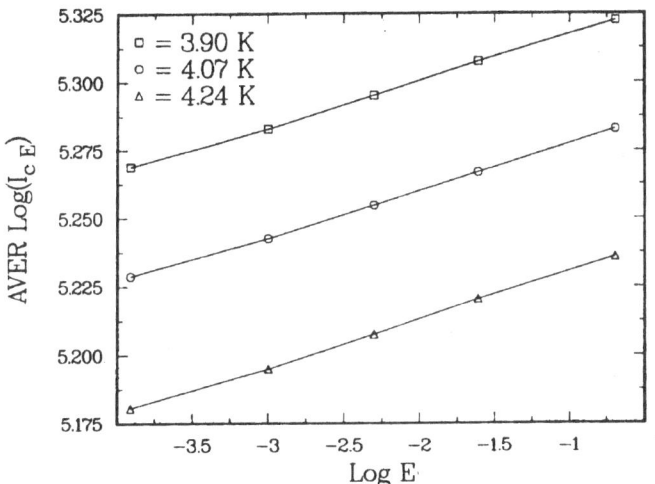

Figure 15b. Dependence of I_c on electric fields from 0.02 to 0.5 µV/cm at 4 T; natural logarithm of I_c and electric field are plotted.

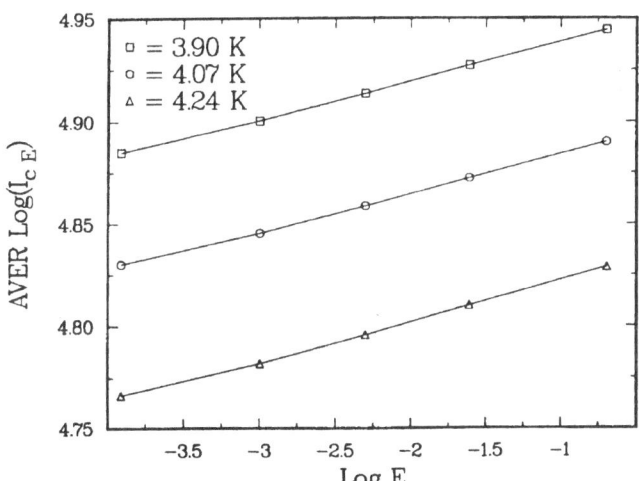

Figure 15c. Dependence of I_c on electric fields from 0.02 to 0.5 µV/cm at 6 T; natural logarithm of I_c and electric field are plotted.

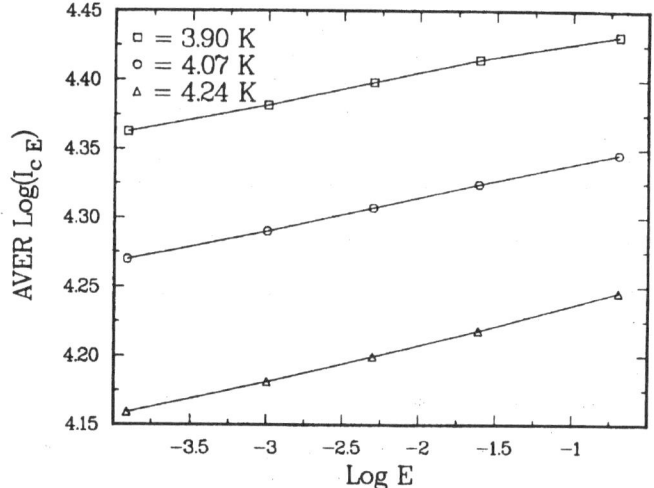

Figure 15d. Dependence of I_c on electric fields from 0.02 to 0.5 µV/cm at 8 T; natural logarithm of I_c and electric field are plotted.

The effects of mechanical and thermal cycling of this SRM cannot be certified for every usage. Examples of mechanical cycling contingencies that would have a cumulative effect are: solder interfaces concentrating the stress of handling; bending in a different orientation; and specimen fatigue. Examples of thermal cycling contingencies that would have a cumulative effect are: the likely automatic repotting of a specimen if petroleum jelly is used; the specimen working up or out of holder grooves; and copper resistivity increased by fatigue. Given the economic reality, a user may wish to reuse a single specimen to perform routine tests. A user may choose to leave a specimen mounted on the holder so that the mechanical cycling factor is minimized. It is conceivable that a user could demonstrate that the particular system and technique employed does not have a cumulative thermal or mechanical cycling effect. In this case, the specimen still has utility and could be used for routine tests. Occasional checks could be performed as needed with a certified specimen. This practice is not discouraged as long as it is clear that this usage cannot be certified.

Related documents and data are included as appendices for the purpose of additional information. Appendix A is a list of the physical parameters of the conductor chosen for the SRM. Appendix B is a copy of the SRM certificate. Appendix C is a copy of the SRM publicity sheet.

9. SUMMARY

This report reviews the selection and certification of a Standard Reference Material (SRM) for the measurement of superconducting critical current. This SRM is intended to provide a means for checking the performance of measurement systems used in the commerce and technology of superconductors.

The critical current was measured using a 2 cm voltage tap separation on a coil specimen holder with a 3.18 cm diameter (specimen bend diameter of about 3.23 cm) and a two turn per centimeter pitch length. The general technique used to measure the critical current was the American Society for Testing and Materials standard test method B714-82 [1]; however the variables were held to much tighter limits in order to get a lower total uncertainty including the sample inhomogeneity. The acquisition and analysis of the raw data was carefully developed to measure the critical current with accuracy and precision. The limits to systematic and random errors of the principal variables that effect the critical current (current, electric field, magnetic field, temperature and strain) were estimated (see table 1). Other concomitant variables and effects considered in the development and certification of this SRM were: voltage filtering and response time; current ramp rate; liquid helium hydrostatic head and stratification; inductive and thermoelectric voltage; magnetic field of the specimen coil; and winding tension.

Preliminary screening measurements were performed on each of five candidate SRM samples. Two of these conductors displayed short- or long-range inhomogeneity that made them seem unfit for use as an SRM. One of the remaining conductors was eliminated becaused the length delivered was considered too short. One of the two candidates left had the lowest copper-to-superconductor ratio, so it was eliminated in favor of the other. Further measurements and tests were made on the conductor chosen to be the SRM.

The conductor designated as SRM 1457 was wound onto 500 distribution spools, each with approximately 2.2 m of wire. Nine of these spools were selected at nearly equal distances along the whole length of wire, including the spools at each end. Critical current measurements on the sample spools were obtained at 36 combinations of three factors affecting critical current: magnetic field, temperature, and electric field.

Although there were no obvious trends along the length of wire, substantial variation in critical current was evident, especially at lower magnetic fields. These variations were associated with material variability, or inhomogeneity, of the wire, and were incorporated into a statistical model that was derived from an empirical equation for the dependence of critical current on temperature and electric field at any given magnetic field.

SRM 1457 has been certified at magnetic fields 2, 4, 6, and 8 T for temperatures from 3.90 to 4.24 K and electric field criteria from 0.05 to 0.2 μV/cm. Because material variability could not be ignored, the uncertainty in the certified values of critical current (excluding systematic errors) is a statistical tolerance interval. The resulting tolerance limits allow for inhomogeneity by estimating limits for the critical current of individual spools, rather than limits on the average critical current of all spools.

The statistical tolerance limits and estimated systemic error have been combined to give a total uncertainty on the certified values. The final estimated uncertainties are no greater than 2.57% of the reported critical current at any of the four magnetic fields.

10. REFERENCES

1. Standard Test Method for D-C Critical Current of Composite Superconductors, Annual Book of ASTM Standards, ASTM B714-82, Part 02.03, pp. 595-98, American Society for Testing and Materials, Philadelphia, PA (1983).

2. J. D. Cali, T. W. Mears, R. E. Michaelis, W. P. Reed, R. W. Seward, C. L. Stanley, H. T. Yolken, and H. H. Ku, "The Role of Standard Reference Materials in Measurement Systems," NBS Monograph 148, National Bureau of Standards, Washington, D.C. (1975) 54 pp.

3. L. F. Goodrich and F. R. Fickett, Critical current measurements: A compendium of experimental results, Cryogenics 22:225-41 (1982).

4. L. F. Goodrich, J. W. Ekin, and F. R. Fickett, Effect of twist pitch on short-sample V-I characteristics of multifilamentary superconductors, Adv. Cryog. Eng. 28:571-80 (1982).

5. M. Durieux, D. N. Astrov, W. R. G. Kemp, and C. A. Swenson, The deviation and development of the 1976 provisional 0.5 K to 30 K temperature scale, Metrologia 15:57-63 (1979).

6. M. Durieux, W. R. G. Kemp, C. A. Swenson, and D. N. Astrov, The 1976 provisional 0.5 K to 30 K temperature scale, Metrologia 15:65-68 (1979).

7. M. H. Edlow and H. H. Plumb, Reproducibilities of carbon and germanium thermometers at 4.2°K, Adv. Cryog. Eng. 6:542-547 (1961).

8. H. H. Sample and L. G. Rubin, Magnetic field induced temperature changes in cryogenic liquids: N_2, Ar, and He^4, Cryogenics 18:223-229 (1978).

9. J. W. Ekin, Strain scaling law and the prediction of uniaxial and bending strain effects in multifilamentary superconductors, in: "Filamentary A15 Superconductors," M. Suenaga and A. F. Clark, eds., Plenum Press, New York, p. 187 (1980).

10. J. W. Ekin, M. B. Kasen, D. T. Read, R. E. Schramm, R. L. Tobler, and A. F. Clark, "Materials Studies for Superconducting Machinery Coil Composites," NBSIR 80-1633, National Bureau of Standards, Boulder, Colorado, pp.89-115 (1979).

11. W. J. Dixon and M. B. Brown, eds., BMDP-79: Biomedical computer programs P-series, Berkeley: University of California Press, Berkeley, California (1979).

12. A. Weissberg and G. H. Beatty, Tables of tolerance - limit factors for normal distributions, Technometrics 2:483-500 (1960).

13. F. A. Graybill, Theory and application of the linear model, Duxbury Press, North Scituate, Massachusetts (1976) p. 642.

14. H. H. Ku, "Notes on the use of propagation of error formulas," NBS Special Publication 300, Vol. I, National Bureau of Standards, Washington, DC (1969).

ACKNOWLEDGMENTS

The help and support of the Superconductors and Magnetic Materials Group, especially F. R. Fickett, J. C. Brauch, and M. S. Allen, is gratefully acknowledged. Assistance of the U.S. wire manufacturers in selecting homogeneous conductors is also appreciated. A special thanks is extended to V. L. Grulke for preparing this manuscript.

APPENDIX A

PHYSICAL PARAMETERS OF SRM

resin type insulation
bare wire diameter of 0.51 mm
180 niobium-titanium filaments
filament diameter approximately 23 μm
1.8 to 1 copper-to-superconductor ratio
filament twist pitch approximately 0.79 twist/cm

APPENDIX B

Certificate

Standard Reference Material 1457

Superconducting Critical Current - NbTi Wire

This Standard Reference Material (SRM) is intended to provide a means for checking the performance of measurement systems used in superconductor technology. This SRM consists of 2.2 m of a multifilamentary niobium titanium, copper stabilized superconducting wire wound in a single layer onto a spool with a core diameter of 8.7 cm. Critical current (I_c) for the SRM is certified over a range of magnetic fields, temperatures, and electric field criteria.

Measurement Technique: Adherence to the precautions given here, together with the American Society for Testing and Materials (ASTM) standard test method B714-82 [1], are necessary and sufficient for a valid certification. Measurements for certifying SRM 1457 were obtained on a coil of diameter 3.18 cm with a voltage tap separation of 2 cm. The critical current is defined as the average of the values measured with increasing and decreasing current.

Certified Critical Currents: The certified critical currents in amperes at 4.2 K for an electric field criterion of 0.2 µV/cm are given in the table. At respective magnetic fields, critical currents of this SRM can be calculated for temperatures from 3.90 to 4.24 K and electric field criteria from 0.05 to 0.2 µV/cm using:

$$I_c(T, E) = I_c(4.2, 0.2) \cdot \{\exp[A(4.2 - T) + B(4.2 - T)^2]\} \cdot [E/0.2]^C$$

where $I_c(4.2, 0.2)$ and the coefficients A, B, and C are given in the table. Critical currents for SRM 1457 were derived from an empirical equation for the dependence of critical current on temperature (T) and electric field criterion (E): $\ln(I_c) = \ln(I_r) + A(T_r - T) + B(T_r - T)^2 + C\ln(E/E_r)$. In this equation, $\ln(I_c)$ is the natural logarithm of the critical current and I_r is the critical current at the reference temperature, T_r, and reference electric field criterion, E_r. The experimental data at each magnetic field were fitted by a maximum likelihood procedure using a statistical model [2] that combines the empirical expression above with terms that allow for material variability (inhomogeneity) among the spools of wire.

Statistical design and data analysis were provided by D. F. Vecchia of the Statistical Engineering Division. Measurements for certification of SRM 1457 were coordinated by L. F. Goodrich. The measurements leading to the development and certification of SRM 1457 were performed by L. F. Goodrich, E. S. Pittman, and A. F. Clark of the Electromagnetic Technology Division.

The technical support aspects involved in the preparation, certification, and issuance of this Standard Reference Material were coordinated through the Office of Standard Reference Materials by R. K. Kirby.

Washington, D.C. 20234   Stanley D. Rasberry, Chief
February 24, 1984   Office of Standard Reference Materials

Certified Value of Critical Current (I_c) at 4.2 K and 0.2 μV/cm and
Coefficients for Temperature and Electric Field Extrapolation.

Magnetic Field (T)	Critical Current (A)	Total Uncertainty (%)	Coefficients for Extrapolation		
			A	B	C
2.000	293.30*	±2.57	0.218625	−0.04755	0.0172089
4.000	187.38	±2.01	0.266361	−0.04682	0.0176600
6.000	124.72	±1.71	0.369479	−0.10488	0.0194218
8.000	69.72	±1.97	0.649242	−0.27906	0.0248311

*Extra digits are provided for accurate extrapolation.

Interpretation of Uncertainty: The uncertainty of a certified critical current at each magnetic field is the sum of an estimated systematic error and statistical tolerance limits computed from the experimental data. The total uncertainty is expressed as percent error in I_c and does not change for extrapolated critical currents over the allowable range of temperature and electric field criteria.

The statistical tolerance limits were constructed so that they should include 99 percent of critical current measurements with probability 0.95. The resulting tolerance interval (and total uncertainty) is valid for a single measurement on any given spool that is made as directed on a coil of diameter 3.18 cm with a voltage tap separation of 2 cm.

Precautions:

1) This SRM should be carefully handled and stored to protect it against physical damage such as: excessive bending, scraping, and other deformation. Any excessive physical damage will invalidate the certification.

2) On each spool, the twisted wire ends and an additional 2 cm on each end of the spool core should be discarded. These sections of the wire are not certified.

3) This certification is invalidated if this SRM is mechanically cycled by demounting and remounting on a specimen holder. Mechanical cycling can concentrate handling stress, which would lower I_c in the stressed regions of the conductor.

4) This certificate is based on a slow cooling of the specimen mounted on a G-11 tube (circumferential fiber direction) by gas heat exchange with a liquid nitrogen precooled dewar [2]. For a valid certification, the specimen must be measured on a suitable specimen holder [1]. The effect of a rapid cooling by immersion into liquid nitrogen or liquid helium can change I_c owing to dynamic differential thermal contraction. For rapid (immersive) cooling, 0.25% must be added to the total uncertainty even if a suitable specimen holder is used. It is conceivable that a user could demonstrate that the particular system and technique employed does not have a cumulative thermal or mechanical cycling effect. In this case, the specimen still has utility but, this SRM can not be certified beyond one thermal cycle.

51

5) This certification is only valid for a zero-to-I_c ramp time in the range of 30 to 300 seconds for all magnetic fields [2]. Also, the certification was based on the assumption that voltage filtering and instrumentation response times contribute negligible error to the measured value of I_c. A nonnegligible effect can be removed by averaging the I_c values measured with increasing and decreasing current at a constant ramp rate. A nonnegligible effect must be removed for a valid certification.

6) A chemical wire stripping compound should be used to remove the insulation from this SRM. A phenol/methylene chloride wire stripping compound was found to adequately remove this insulation.

7) If the specimen temperature exceeds 250°C, the certification is invalidated. The current and voltage contacts should be soldered carefully to avoid overheating and physical damage. If a specimen enters the normal state (quenches) while carrying a high current density, it could melt within a few seconds. An adequate quench protection circuit may be necessary [1]. A typical current shutdown time of 10 ms is adequate.

8) For a voltage tap separation of more than 2 cm, the uncertainty in I_c should be less. For a voltage tap separation of less than 2 cm, the uncertainty in I_c may be more. This certification is only valid for a voltage tap separation greater than or equal to 2 cm.

9) If this SRM is measured with a bend diameter other than 3.23 cm (coil of diameter 3.18 cm), the results may be different. Uniaxial strain data was used to determine the expected upper limits to the bending strain effect. For bending diameters from infinity (straight) to 1.6 cm, the certified critical current values can be used only if the following amount is added to the total uncertainty:

$$G \cdot \left| 1 - (3.23/d)^2 \right|,$$

where d is the bend diameter in centimeters and G=1.10, 1.20, 1.36, and 1.70% at 2, 4, 6, and 8 T, respectively.

Non-certified Values at Other Criteria: Critical current measurements were made on the sample spools at electric field criteria 0.02 and 0.5 µV/cm. However, SRM 1457 is not certified to the extended range of electric field because measurements at the additional criteria did not conform to the required measurement procedure. A comparison of observed critical currents to values extrapolated from the certifying equation is given in reference 2 for information only. Most of the measurements were within the range computed using tolerance limits that are only valid between 0.05 and 0.2 µV/cm.

References:

1. Standard Test Method for D-C Critical Current of Composite Superconductors, Annual Book of ASTM Standards, ASTM B714-82, Part 2.03, pp. 595-98, American Society for Testing and Materials, Philadelphia, PA (1983).
2. Goodrich, L. F., Vecchia, D. F., Pittman, E. S., Ekin, J. W., and Clark, A. F., Critical Current Measurements on an NbTi Superconducting Wire Standard Reference Material, NBS Special Publication 260- (1984).

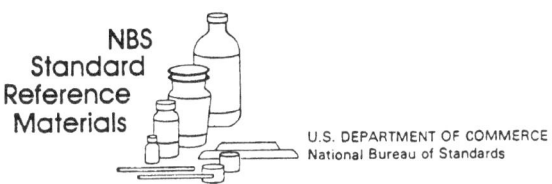

NBS
Standard
Reference
Materials

U.S. DEPARTMENT OF COMMERCE
National Bureau of Standards

Standard Reference Material 1457

Superconducting Critical Current

NbTi Wire

Winter 1984

The NBS Office of Standard Reference Materials announces the availability of the first superconducting wire for critical current measurements as a Standard Reference Material (SRM). It is intended to provide a means for testing the performance of measurement systems used in the development of superconductors. This SRM consists of approximately 2.2 m of a multifilamentary niobium-titanium, copper stabilized, superconducting wire wound in a single layer onto a spool with a core diameter of 8.7 cm.

SRM 1457 should prove valuable in determining the overall accuracy of a critical current measurement system that is dependent on numerous variables and effects that can make this seemingly easy measurement very difficult.

The critical current for SRM 1457 has been certified at magnetic fields of 2, 4, 6, and 8 T, for temperatures from 3.90 to 4.24 K, and electric field criteria from 0.05 to 0.2 μV/cm.

An effort was made to keep the use of this SRM as unrestricted as possible. The precautions listed on the certificate, together with the American Society for Testing and Materials (ASTM) Standard Test Method (B714-82), are sufficient for a valid user measurement technique. Some deviations in testing technique, from the method on which the certification was based, were accommodated by increasing the total uncertainty of the certified critical current. The deviations that are allowed, and the ones that are not allowed, are identified in the precautions sections of the certificate.

SRM 1457 may be purchased from the Office of Standard Reference Materials, Room B311, Chemistry Building, National Bureau of Standards, Washington, DC 20234, at a price of $219.

0384

3. Publication

Septembe:

4. TITLE AND SUBTITLE   Standard Reference Materials:

Critical Current Measurements on an NbTi Superconducting Wire
Standard Reference Material

6. PERFORMING ORGANIZATION (If joint or other than NBS, see instructions)

NATIONAL BUREAU OF STANDARDS
DEPARTMENT OF COMMERCE
BOULDER, CO 80303

9. SPONSORING ORGANIZATION NAME AND COMPLETE ADDRESS (Street, City, State, ZIP)

Office of Standard Reference Materials
National Measurement Laboratory
National Bureau of Standards
Gaithersburg, MD 20899

Office of Fusion Energy
and Office of High Energy Physics
Department of Energy
Washington, DC 20545

10. SUPPLEMENTARY NOTES

Library of Congress Catalog Card Number: 84-601108

☐ Document describes a computer program; SF-I85, FIPS Software Summary, is attached.

11. ABSTRACT (A 200-word or less factual summary of most significant information. If document includes a significant
bibliography or literature survey, mention it here)

This report reviews the selection and certification by NBS of a Standard
Reference Material (SRM) for the measurement of superconducting critical current.
Procedures for preparing and measuring five candidate conductors are described.
Evaluation criteria are discussed by which one of the five conductors was selected
for the critical current SRM. The designated superconducting wire, SRM 1457, has
been subdivided and wound onto 500 spools for distribution. Certified critical
current measurements were made on a sample of these spools. Material variability,
or inhomogeneity, along the whole wire is included in a statistical model based
on the dependence of critical current on temperature and electric field. Critical
currents for SRM 1457 are certified at magnetic fields of 2, 4, 6, and 8 T for
temperatures from 3.90 to 4.24 K and electric field criteria from 0.05 to 0.2 μV/cm.
Statistical tolerance limits and estimated systematic errors are combined to give an
overall uncertainty in the certified values. The total uncertainty is no greater
than 2.57 percent of the reported critical current at any of the four magnetic
fields.

13. AVAILABILITY

☒ Unlimited
☐ For Official Distribution. Do Not Release to NTIS
☒ Order From Superintendent of Documents, U.S. Government Printing Office, Washington, D.C.
20402.

☐ Order From National Technical Information Service (NTIS), Springfield, VA. 22161

NBS Technical Publications

Periodicals

Journal of Research—The Journal of Research of the National Bureau of Standards reports NBS research and development in those disciplines of the physical and engineering sciences in which the Bureau is active. These include physics, chemistry, engineering, mathematics, and computer sciences. Papers cover a broad range of subjects, with major emphasis on measurement methodology and the basic technology underlying standardization. Also included from time to time are survey articles on topics closely related to the Bureau's technical and scientific programs. As a special service to subscribers each issue contains complete citations to all recent Bureau publications in both NBS and non-NBS media. Issued six times a year.

Nonperiodicals

Monographs—Major contributions to the technical literature on various subjects related to the Bureau's scientific and technical activities.

Handbooks—Recommended codes of engineering and industrial practice (including safety codes) developed in cooperation with interested industries, professional organizations, and regulatory bodies.

Special Publications—Include proceedings of conferences sponsored by NBS, NBS annual reports, and other special publications appropriate to this grouping such as wall charts, pocket cards, and bibliographies.

Applied Mathematics Series—Mathematical tables, manuals, and studies of special interest to physicists, engineers, chemists, biologists, mathematicians, computer programmers, and others engaged in scientific and technical work.

National Standard Reference Data Series—Provides quantitative data on the physical and chemical properties of materials, compiled from the world's literature and critically evaluated. Developed under a worldwide program coordinated by NBS under the authority of the National Standard Data Act (Public Law 90-396).
NOTE: The Journal of Physical and Chemical Reference Data (JPCRD) is published quarterly for NBS by the American Chemical Society (ACS) and the American Institute of Physics (AIP). Subscriptions, reprints, and supplements are available from ACS, 1155 Sixteenth St., NW, Washington, DC 20056.

Building Science Series—Disseminates technical information developed at the Bureau on building materials, components, systems, and whole structures. The series presents research results, test methods, and performance criteria related to the structural and environmental functions and the durability and safety characteristics of building elements and systems.

Technical Notes—Studies or reports which are complete in themselves but restrictive in their treatment of a subject. Analogous to monographs but not so comprehensive in scope or definitive in treatment of the subject area. Often serve as a vehicle for final reports of work performed at NBS under the sponsorship of other government agencies.

Voluntary Product Standards—Developed under procedures published by the Department of Commerce in Part 10, Title 15, of the Code of Federal Regulations. The standards establish nationally recognized requirements for products, and provide all concerned interests with a basis for common understanding of the characteristics of the products. NBS administers this program as a supplement to the activities of the private sector standardizing organizations.

Consumer Information Series—Practical information, based on NBS research and experience, covering areas of interest to the consumer. Easily understandable language and illustrations provide useful background knowledge for shopping in today's technological marketplace.
Order the above NBS publications from: Superintendent of Documents, Government Printing Office, Washington, DC 20402.
Order the following NBS publications—FIPS and NBSIR's—from the National Technical Information Service, Springfield, VA 22161.

Federal Information Processing Standards Publications (FIPS PUB)—Publications in this series collectively constitute the Federal Information Processing Standards Register. The Register serves as the official source of information in the Federal Government regarding standards issued by NBS pursuant to the Federal Property and Administrative Services Act of 1949 as amended, Public Law 89-306 (79 Stat. 1127), and as implemented by Executive Order 11717 (38 FR 12315, dated May 11, 1973) and Part 6 of Title 15 CFR (Code of Federal Regulations).

CPSIA information can be obtained
at www.ICGtesting.com
Printed in the USA
BVHW041347280119
538843BV00005B/240/P